Civil Engineering Hydraulics

Civil Engineering Hydraulics

Contributors

Xi Zhang, Rob Jeffrey et al.

AURIS
Reference

www.aurisreference.com

Civil Engineering Hydraulics

Contributors: Xi Zhang, Rob Jeffrey et al.

Published by Auris Reference Limited

www.aurisreference.com

United Kingdom

Civil Engineering Hydraulics

ISBN: 978-1-78154-909-4

British Library Cataloguing in Publication Data
A CIP record for this book is available from the British Library

Printed in the United Kingdom

Exclusively distributed by CBS Publishers & Distributors Pvt. Ltd.

Sales & Distribution Rights only for India, Pakistan, Bangladesh, Sri Lanka, Nepal and Bhutan.This book is not to be sold outside these territories.

Contents

List of Abbreviations

BEM	Boundary Element Method
BMP	Best Management Practices
CFD	Computational Fluid Dynamics
DEM	Digital Elevation Model
DL	Detection Limit
ELWNR	Ebinur Lake Wetland Nature Reserve
EPA	Environmental Protection Agency
EPM	Equivalent Porous Media
FEM	Finite Element Method
GDI	Growth dominance Index
G-M	Genuchten-Mualem
GSD	Grain-size Distribution
HME	Hydraulic Model Experiment
HR	hydraulic Redistribution
L&D	Lock and Dam
LCR	Lower Columbia River
LGRB	Landesamt für Geologie, Rohstoffe und Bergbau
LID	Low Impact Development
LSMs	large-scale Motions
LSPIV	Large-Scale Particle Image Velocimetry
MAE	Mean absolute error
MCL	Maximum Contamination Level
NME	Normalized Mean Error
NOAA	National Oceanic Atmospheric Administration
NPL	National Priorities List
PIV	Particle Image Velocimetry
PRASA	Puerto Rico Aqueduct and Sewer Authority
PSD	Pore-size Distribution
REV	Representative elementary volumes
RHC	Hydraulic Conductivity
RMSE	Root Mean Square Error
SCPE	Single continuum Porous Equivalent Approach
SUDS	Sustainable Urban Drainage System
SWCC	Soil Water Characteristic Curve
SWD	Soil-water Diffusivity
SWWRP	System Wide Water Resources Program
TCE	Trichloroethylene
USACE	US Army Corps of Engineers
USGS	U. S. Geological Survey
USGS	United States Geological Survey
VITA	variable integration time average
VOCs	Volatile Organic Compounds
WD	Water Divide
WSUD	Water Sensitive Urban Design

List of Contributors

Xi Zhang
CSIRO Earth Science and Resource Engineering, Melbourne, Australia

Rob Jeffrey
CSIRO Earth Science and Resource Engineering, Melbourne, Australia

Marcelo Gomes Miguez
Federal University of Rio de Janeiro Brazil

Aline Pires Veról
Federal University of Rio de Janeiro Brazil

Paulo Roberto Ferreira Carneiro
Federal University of Rio de Janeiro Brazil

Sanghwa Jung
Department of Water Resources Research & Environment Research, Korea Institute of Construction Technology, Ilsan, Korea

Joongu Kang
Department of Water Resources Research & Environment Research, Korea Institute of Construction Technology, Ilsan, Korea

Il Hong
Department of Water Resources Research & Environment Research, Korea Institute of Construction Technology, Ilsan, Korea

Hongkoo Yeo
Department of Water Resources Research & Environment Research, Korea Institute of Construction Technology, Ilsan, Korea

Xiao-Dong Yang
College of Ecological and Environmental Sciences, East China Normal University, Shanghai, China
Tiantong National Forest Ecosystem Observations and Research Station, Chinese National Ecosystem Observation and Research Network, Ningbo, China

Xue-Ni Zhang
Institute of Resources and Environment Science, Xinjiang University, Urumqi,

China
Xinjiang Key Laboratory of Oasis Ecology, Ministry of Education, Urumqi, China

Guang-Hui Lv
Institute of Resources and Environment Science, Xinjiang University, Urumqi, China
Xinjiang Key Laboratory of Oasis Ecology, Ministry of Education, Urumqi, China

Arshad Ali
College of Ecological and Environmental Sciences, East China Normal University, Shanghai, China
Tiantong National Forest Ecosystem Observations and Research Station, Chinese National Ecosystem Observation and Research Network, Ningbo, China

Giuliana Zanchi
Department of Physical Geography and Ecosystem Science, Lund University, Sölvegatan 12, SE-223 62 Lund, Sweden

Salim Belyazid
Department of Physical Geography, Stockholm University, Svante Arrhenius väg 8, SE-114 18 Stockholm, Sweden

Cecilia Akselsson
Department of Physical Geography and Ecosystem Science, Lund University, Sölvegatan 12, SE-223 62 Lund, Sweden

Lin Yu
Centre for Environmental and Climate Research, Lund University, Sölvegatan 37, SE-223 62 Lund, Sweden

Kevin Bishop
Department of Aquatic Sciences and Assessment, SLU, Uppsala, Sweden

Stephan J. Köhler
Department of Aquatic Sciences and Assessment, SLU, Uppsala, Sweden

Harald Grip
Department of Forest Ecology and Management, SLU, Surbrunnsgatan 4, SE-114 27 Stockholm, Sweden

Ronald J. Adrian
Ira A. Fulton Professor, School for Engineering of Matter, Transport and Energy, Arizona State University,
Tempe, AZ 85287, USA

Willi H. Hager
IAHR Honorary Member, Professor, VAW, ETH Zurich, CH-8093 Zürich, Switzerland

Robert M. Boes
IAHR Member, Professor and Director, VAW, ETH Zurich, CH-8093 Zürich, Switzerland

Yongfu Xu
Department of Civil Engineering, Shanghai Jiao Tong University, Shanghai, China

Reza Ghasemizadeh
Department of Civil and Environmental Engineering, Northeastern University, Boston, Massachusetts 02115, United States of America

Xue Yu
Department of Civil and Environmental Engineering, Northeastern University, Boston, Massachusetts 02115, United States of America

Christoph Butscher
Department of Engineering Geology, Institute of Applied Geosciences, Karlsruhe Institute of Technology, 76131, Karlsruhe, Germany

Ferdi Hellweger
Department of Civil and Environmental Engineering, Northeastern University, Boston, Massachusetts 02115, United States of America

Ingrid Padilla
Department of Civil Engineering and Surveying, University of Puerto Rico, Mayaguez, Puerto Rico 00682, United States of America

Akram Alshawabkeh
Department of Civil and Environmental Engineering, Northeastern University, Boston, Massachusetts 02115, United States of America

Gaurav Savant
Research Water Resources Engineer, Dynamic Solutions LLC, 6421 Deane Hill Drive, Suite 1, TN 37919, USA

Tate O. McAlpin
Research Physicist, Engineer Research and Development Center, US Army Corps of Engineers, 3909 Halls Ferry Road, MS 39180, USA

S. Oehlmann
Geoscience Center, University of Göttingen, Germany

T. Geyer
Geoscience Center, University of Göttingen, Germany

T. Licha
Geoscience Center, University of Göttingen, Germany

S. Birk
Institute for Earth Sciences, University of Graz, Graz, Austria

Preface

Hydrology deals with the origin, occurrence, circulation, distribution, the physical and chemical properties of water and its interaction with living organisms. As a sub-discipline of civil engineering, hydraulic engineering is concerned with the flow and conveyance of fluids, principally water and sewage. The text *Civil Engineering Hydraulics* provides a succinct introduction to the theory of civil engineering hydraulics. The purpose of first chapter is to provide some initial results for hydraulic fracture growth through a natural fracture network. An integrated approach, combining hydraulic engineering design, urban land control, and river revitalization aspects for sustainable drainage systems have been proposed in second chapter. Third chapter analyzes the hydraulic features affecting the surrounding stability when installing floating islands and proposes stable floating islands layout in terms of hydraulics based on the experiment results. The objective of fourth chapter is to test for P. euphratica HR and to explain the effect of HR on plants growth condition and species biodiversity maintenance in the Ebinur desert. Fifth chapter presents a hydrology concept developed to include lateral water flow in the biogeochemical model for safe. In sixth chapter, we review the significant advances over the past few decades in the fundamental study of wall turbulence over smooth and rough surfaces, with an emphasis on coherent structures and their role at high Reynolds numbers. A positive outlook into the future of hydraulic structures has been presented in seventh chapter. Eighth chapter deals with unsaturated hydraulic conductivity of fractal-textured soils. Last chapter evaluates the application of the equivalent porous media (EPM) approach to simulate groundwater hydraulics and contaminant transport in karst aquifers using an example from the North Coast limestone aquifer system in Puerto Rico.

Chapter 1

DEVELOPMENT OF FRACTURE NETWORKS THROUGH HYDRAULIC FRACTURE GROWTH IN NATURALLY FRACTURED RESERVOIRS

Xi Zhang and Rob Jeffrey

CSIRO Earth Science and Resource Engineering, Melbourne, Australia

ABSTRACT

A 2-D numerical study was carried out, using a fully coupled rock deformation and fluid flow hydraulic fracturing model, on fracture network formation by advancing, widening and interconnecting discrete natural fractures in a low-permeability rock, some of which are small enough to be considered as a flaw that acts as a fracture seed. The model also includes fractures connecting into one another to form a single hydraulic fracture. In contrast to previous fracture network models, fracture extension and fluid flow behavior, frictional slip, and fracture interaction are all explicitly addressed in this model. Incompressible Newtonian fluid is injected at a constant total rate into fractures to study viscous fluid effects on the network formation. The algorithm for flow division and coalescence is validated through some examples.

Numerical results show that the incremental crack propagation that connects isolated natural fracture sets depends on the current stress state and the fracture arrangement. The newly created connecting fracture segments increase local conductivity since they are oriented along a path that is easier to open when pressurized by fluid and provide a new path for fluid flow. However the hydraulic fracture growth process is regarded by some of the resulting geometric changes such as intersections and offsets, and the growth-induced sliding that can impose a barrier to further fracture growth and fluid flow into parts of the network. Such barriers may eventually result in a fracture branch initiating and growing that results in a relatively shorter and more conductive path through a fracture network zone.

We consider a specific fracture arrangement consisting of around 20 conductive pre-existing fractures to study the effective behavior of the hydraulic fracture growth through a natural fracture network. Mechanical responses have been studied for two different fracture and flow scenarios depending on the fluid entry details: one fracture system assumes each of four entry fractures has one quarter of the total injection rate and the other system is defined to maintain the influx rates into each inlet fracture so that the pressure across all four inlet fractures is equal (but not necessarily constant in time). For the latter case, a preferential flow pathway is developed as a result of hydraulic fracture growth and the overall permeability of the fracture system increases rapidly after this hydraulic fracture path develops. The former injection condition results in development of more evenly distributed advancing fractures that provide a more homogeneous flow pattern.

INTRODUCTION

The understanding of fluid movement through fractured rock masses is essential to improving the success of reservoir stimulations for energy resources such as shale gas and geothermal energy and for stimulation by hydraulic fracturing of any naturally fractured rock mass. As we know, fractures play an important role in the flow of fluid through rock masses by building connected networks that channel flow. These networks develop either through enhanced conductivity of existing fractures or by new fracture growth that connects existing fractures. There are many studies in the literature devoted to characterising the fracture system connectivity in relation to fracture orientation, size and conductivity (see the comprehensive review [1]). Besides these static factors, the time dependent evolution of fluid pressure and stress states can generate different fracture patterns that act to assist or inhibit further fluid flow, in particular forming a preferential fluid pathway in the presence of viscous fluid flow [2, 3]. Under some circumstances, the fracture growth driven by pressurized fluids can propogate a hydraulic fracture to connect two isolated fracture clusters. Cross-cutting fractures in the connected region are filled with fluid and pressurized but may open or not. Therefore fluid movement in a fractured rock mass will involve both new fracture growth and permeability enhancement of existing fractures. Clearly, using an equivalent porous continuum model to represent fluid flow and fracture growth would be inaccurate, especially when the flow is dominated by hydraulic fracture processes. The approach here is to study the full coupled process in order to determine the parameters that control fracture development. Simplifications and averaging methods can eventually be realistically employed without degrading the ability to predict fracture growth.

Discrete fracture models, so named because these models treat fractures as discrete entities, are applicable whenever the process involves fractures growth and flow where details such as opening, shearing or growth of the fractures are being studied. If many fractures are considered, these discrete models become computationally demanding. Such a system is very heterogeneous and localized in both fracture growth and fluid flow. Early numerical models treated the hydraulic fractures as single planar fractures and did not consider fracture interaction [4-6]. The emergent behaviours associated with fracture propagation under a tensile displacement boundary condition have been described as straight paths using a subcritical failure criterion and a propagation speed exponent. Recently, the effect of curving fractures on fluid flow in both the fractures and the matrix under a tensile displacement boundary condition has been considered [7,8]. However, the pressure distributions used inside the fractures are either uniform [7] or based on the steady-state solution for flow in a porous medium [8]. The effect of fluid viscosity on pressure distribution has therefore not been considered in these fracture network models. Moreover, some simplifications in the fracture geometry changes, such as the details of interactions at intersections and in regions where fractures close, are used in these models and these simplications affect fluid flux distribution. Uncoupling of deformation and fluid flow, as used in some models, limits the application of the results obtained.

Fracture models that do not consider flow in the matrix, are applicable to low-permeability rocks [9]. In general, the fracture aperture is very narrow and would be different in each fracture branch intersecting at the same point, and these differences are expected to be significant in their effect on fluid movement through the intersections. In contrast to the stress-driven uniform fracture nucleation in porous rocks subject to a tensile loading environment, the propagation of a hydraulic fracture through a network of pre-exsiting fractures is dependent on local stress states around fracture tips, which in turn depends on the fluid flow and pressure along the non-planar hydraulic fracture path. Under most circumstances, one dominant hydraulic fracture is generated and the entire injected fluid rate is carried to the outlet through this preferential flow path, while shorter natural fractures that are intersected by this main fracture remain closed or act as dead end branches [9]. Models that include the coupling of rock deformation and viscous fluid flow provide a means of studying the fracture development and the evolution of distinct preferential flow paths and development of a dominant fracture.

At intersections of two or more fractures, the kinematic deformation transfer between slip and opening of the fractures can induce additional fracture aperture changes. In addition, the viscous effect becomes stronger for narrow channel widths, which are commonly associated with intersections and offsets. The importance of tracking the details of fracture geometry lies in the fact that although the pressure losses may diminish after a long time, initial fracture geometry details may strongly affect the final fracture patterns. Studies that neglect viscous fluid effects, by using uniform pressure or steady-state transport and deformation models, will, in many cases miscalculate the stresses and flow rates, thus producing incorrect fracture and flow patterns as time-dependent pressure responses are not determined accurately.

Mechanical interaction among fractures has not received sufficient attention in the literature involving discrete fracture models especially for cases involving pressurized viscous fluids sufficient to result in hydraulic fracture growth. Any inaccuracy in the calculated fracture pathway may cause incorrect flux redistribution at intersections, as fluid flow behaviour is strongly dependent on local width. Due to intrinsic complexity of the problem, numerical methods appear to be the only approach able to explicitly solve the nonlinear and nonlocal fracture-fluid and fracture-fracture interaction in such fracture network. The Distinct Element Method (DEM) and the Finite Element Method (FEM) have been used for this purpose [10,11]. However, the fracture pathways are confined along the element edges in classical FEM models and the out-of-plane fracture propagation is difficult to accurately simulated because it requires remeshing. We have developed a Boundary Element Method (BEM) based program for treating this coupled problem [12,13]. The validation of the code has been carried out for various simple cases involving both viscous fluid and uniform pressure. Additionally, the program treats rock deformation and fluid flow as a whole in that the field variables are obtained in a single framework, instead of the one-way coupling scheme as used in Reference [11].

The purpose of this paper is to provide some initial results for hydraulic fracture growth through a natural fracture network. Fracture growth is allowed and is based on a local failure criterion[14] and fracture coalescence can take place to form a path through an existing network of natural fractures. The numerical treatment of fracture coalescence has been detailed in [12,13]. The rock mass is assumed to be impermeable and the fluid flow is confined to occur along the pre-existing or newly created fractures. The fracture nucleation sites are embedded as the pre-existing secondary fractures with small sizes to reflect the tensile strength heterogeneities existing along the fracture surface [15]. For this plane-strain model, the strain in the out-of-plane direction is assumed to be zero and the fracture should be visualised as extending uniformly a significant distance in this direction.

THE MODEL

The basic governing equations and the boundary conditions are provided in our previous work (see References [12,13,16]), dealing with the hydraulic fracture model that fully couples mechanisms of rock deformation and viscous fluid flow. The basis of the model is briefly described here for the sake of completeness.

We allow for the fracture surfaces to be rough and tortuous, which imparts a hydraulic aperture for the closed fracture allowing fluid flow, but causing no stress and deformation changes. Fluid volumetric flux q in a closed fracture segment is described by:

$$q = -\frac{\varpi^3}{\mu'}\frac{\partial p_f}{\partial s}$$

(1)

and in an opened fracture portion it is

$$q = -\frac{(w + \varpi)^3}{\mu'}\frac{\partial p_f}{\partial s}$$

(2)

where $\mu'=12\mu$ with μ being fluid dynamic viscocity. w and ϖ are mechanical opening and hydraulic aperture along the fracture surface and are functions of time and location. The former is determined by the stress condition given below, but the latter obeys an evolution equation linearly proportional to fluid pressure change [13]. It is noted that the initial value of hydraulic aperture is denoted as ϖ_0, which is a reflection of the fracture surface roughness and tortuosity.

Fluid flow in the opened fracture portion is based on the lubrication equation:

$$\frac{\partial(w + \varpi)}{\partial t} = \frac{\partial}{\partial s}\left(\frac{(w + \varpi)^3}{12\mu}\frac{\partial p_f}{\partial s}\right)$$

(3)

but fluid flow in the closed fracture segment is based on the pressure diffusion equation

$$\frac{\partial p_f}{\partial t} - \frac{1}{\chi_1 \mu'}\frac{\partial}{\partial s}(\varpi^2 \frac{\partial p_f}{\partial s}) = 0$$

(4)

where χ_1 is the compressibility of the fracture with units of Pa^{-1} and it is set as $10-810-8Pa^{-1}$ in this paper. When fracture surfaces are separated, the change of the hydraulic aperture ceases, but its contribution to fluid flow is retained as provided in Eq. (3).

The nonlocal elastic equilibrium equations for a system of N fractures are given as:

$$\sigma_n(x,t) - \sigma_1(x) = \sum_{r=1}^{N} \int_0^{l_r} [G_{11}(x,s)w(s,t) + G_{12}(x,s)v(s,t)]ds$$

$$\tau_s(x,t) - \tau_1(x) = \sum_{r=1}^{N} \int_0^{l_r} [G_{21}(x,s)w(s,t) + G_{22}(x,s)v(s,t)]ds$$

$$\tag{5}$$

where the coordinates of a location of a material point are denoted as $x=(x,y)$ in a two-dimensional Cartesian reference framework, t is time and v is the shear displacement discontinuity, ℓ_r is the length of fracture r. σ_n is the normal stress and τ_s is the shear stress carried by the fracture because of its frictional strength, obeying Coulomb's frictional law characterized by the coefficient of friction λ, which limits the shear stress $|\tau_s| \leq \lambda \sigma_n$, that can act in parts of fractures that are in contact, but vanishes along the separated parts. Along the opened fracture portions, we have $\sigma_n = pf$.

In addition, σ_1 and τ_1 are the normal and shear stresses, respectively, along the fracture direction at location x caused by the far-field stress, whose normal and shear components are denoted as σ_{xx}^{∞}, σ_{yy}^{∞} and σ_{xy}^{∞}. G_{ij} are the hypersingular Green's functions, which are proportional to the plane strain Young's modulus, E', where $E' = \dfrac{E}{(1-v^2)}$.

The global mass balance requires

$$\sum \int_0^{l_f} (w + \varpi)ds = Q_0 t$$

$$\tag{6}$$

And the fluid front in the hydraulic fracture will be, in general, not coincident with the fracture tip. The fluid front location is found by using Eqs. (1) and (2) with the following equation

$$\dot{l}_f = q(l_f,t)/w(l_f,t)$$

$$\tag{7}$$

The fracture growth is based on using the maximum hoop stress criterion, with the maximum mixed-mode stress intensity factor reaching a critical value [14]

$$\cos\frac{\Theta}{2}(K_I \cos^2 \frac{\Theta}{2} - \frac{3}{2}K_{II} \sin\Theta) = K_{Ic}$$

$$\tag{8}$$

where K_I and K_{II} are calculated stress intensity factors, K_{Ic} is tensile mode fracture toughness and Θ is the fracture propagation direction relative to the current fracture orientation. The predicted orientation follows the maximum tensile stress direction, and the near-tip stresses are approximated by the analytical LEFM solutions [14].

The problem must be completed by specifying the imposed boundary conditions at the wellbore, that is, the sum of injection rates of hydraulic fractures connected into the wellbore or to the entry zone, should be equal to the given injection rate Q_0. At the fracture tip the displacement discontinuities are zero, $w(\ell_r, t)=0$ and $v(\ell_r, t)=0$. In addition, the entire system is assumed to initially be stationary and unsaturated.

The numerical scheme for the above nonlinear and nonlocal coupled problem has been detailed in our previous paper [16] based on the Displacement Discontinuity Boundary Element Method. The hydraulic fractures and other geological discontinuities like joints and faults, and natural fractures are discretised with constant displacement elements. The model solves the hydraulic fracture problem simultaneously including the effects of viscous fluid flow and coupled rock deformation. The solutions are consistent with existing results. Also, a fracture can intersect another one in its path and a new fracture can be nucleated from a position on the natural fracture. In particular, the new fracture seeds are pre-defined in this paper along some natural fractures. The interested reader is referred to our previous papers for the details on implementation of fracture growth and coalescence [12, 16]. One important check here is the satisfication of fluid mass conservation after fluid branch coalescence, due to redirection of newly-created fractures.

Table 1: Material properties.

Property	Value
Young's modulus E	50 GPa
Poisson's ratio	0.22
Mode I Toughness	1.0 MPa·m0.5MPa·m0.5
Fluid dynamic viscosity	0.01 Pa·sPa·s
Injection rate	0.00002 m2m2/s
Coefficient of friction along fractures	0.8

A SIMPLE TEST PROBLEM

To validate the model on coalescence of fluid flow branches, we present results for one specific case, as shown in Figure 1, where two hydraulic fractures with a spacing 0.8 m driven by the same fluid source (located at the plane with x=-1.0 m on the left of the natural fracture) intersect the natural fracture orthogonally. And the sum of their inflow rates is a constant. A new fracture site is located on the natural fracture that is 2.4 m in length, and it is assumed to be a pre-existing flaw at 0.12 m long. It is anticipated that after intersection, the fluid will invade the natural fracture until it reaches the new fracture in Figure 1(a), which is subject to a less compressive stress. As the new fracture becomes the weakest point to continue crack growth, it will propagate when the fluid pressure reaches a sufficient value, as shown in Figure 1(c). It should be noted that when the middle section of the natural fracture between two hydraulic fractures is filled with fluid, a sudden increase in the pressure occurs as indicated in Figure 1(a) and (b) for two close time steps, in the same way as injection into a closed container. At the time shown in Figure 1(c) the natural fracture appears to be full of fluid, the whole fracture system experiences a similar pressure level, high enough to cause the new fracture to propagate. The material constants used for this problem are listed in Table. 1 if not otherwise specified.

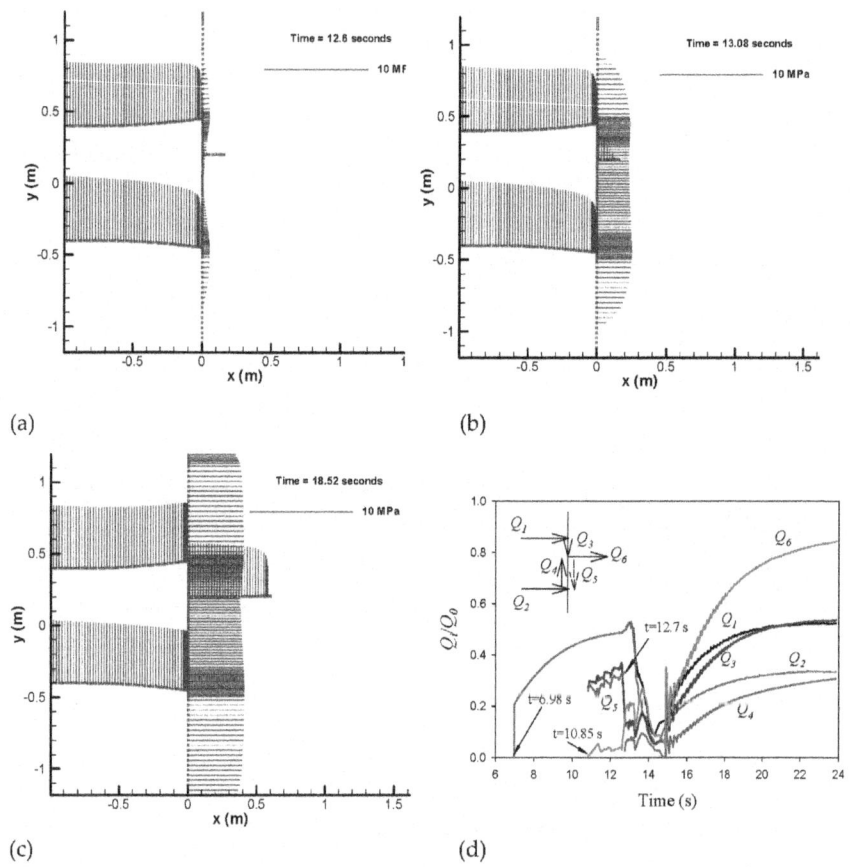

(a)   (b)

(c)   (d)

Figure 1: The pressure profiles at three specific time instants: (a) t=12.6 s, (b) t=13.08s and (c) t=18.5 s and (d) the evolution of influxes around the intersection points and the fracture nucleation site for the case of one natural fracture and two hydraulic fractures that approach the natural fracture at a right angle. The fracture nucleation site on the natural fracture is located at 0.2 m above the mid plane of the two hydraulic fractures.

The applied stress components are σ_{xx}^{∞} =5 MPa, σ_{yy}^{∞} =4 MPa and σ_{xy}^{∞} =0. The initial hydraulic aperture ϖ_0 for the natural fracture is 0.03 mm.

The variations of flow rates into each branch with time are provided in Figure 1(d). The two hydraulic fractures reach the natural fracture at time t=6.98 s and the fluid front reaches the new fracture site at t=10.85 s. In Figure 1(a), the inflow into the new fracture is only from the top hydraulic fractures up to the onset of coalescence of two fluid branches with the natural fracture at t=12.6 s. There is a short-time period where Q_5 occurs before fluid branch coalescence and this flux is represented by Q_4 later on. Subsequently, the value of Q_6 increases rapidly to a higher level (Q_6/Q_0 =0.8). This implies that most of injected fluid is entering the new fracture and this promotes its growth. The fluid rates at the injection fractures all meet the continuity requirement. The larger value of Q_3 compared to Q_4 indicates that the hydraulic fracture closer to the fracture nucleation site is contributing more in fluid flux to sustain the new fracture growth. As for the loss of geometric symmetry, it is interesting to note that the outflux from the top hydraulic fracture is also larger than its counterpart since Q_1>Q_2 clearly shown in Figure 1(d).

RANDOM FRACTURE GEOMETRY

In this section, we present more complex cases where several high-angle joints are defined in a random distribution ahead of hydraulic fractures that grow from a left entry zone (along the plane x=-1 m) toward the right (through the plane x=2 m). The remote stress conditions and some geometric parameters are provided in the caption of Figure 2. The fracture injection occurs into four individual fractures located on the far left, as shown in Figure 2. The fracture segments coloured green denote the initial existing fracture configuration. Each existing fracture consists of a single fracture or of several connected fractures of different sizes and orientations. The hydraulic fractures driven by pressure act to connect these separate fractures to form a conductive path from left to right and the newly created fractures are coloured in red in Figure 2. Two different injection boundary conditions are used in the computations. One is associated with even distribution of the injection rate into the four entry fractures on the left, and the second condition specifies that the pressure at each of these four initial fractures is equal. The latter condition is physically reasonable if we assume a wellbore lies along the y-axis of Figure 2, while the former condition would require isolation and injection into each fracture at the same rate.

Of course, some numerical difficulties exist in the simulation of the process of connecting two intersecting fractures [16]. A mesh sensitivity analysis has been carried out so that the change in fracture orientation does not affect the numerical accuracy as the fracture direction has to be such that the intersecton occurs at the the end of an element on the fracture. A fracture connection event that results in a strong deflection from the calculated direction can influence the subsequent fluid flow. In these cases, the natural fractures must be re-meshed and the job must be run again. In the numerical simulation, the individual element size in the vicinity of intersection point is chosen with care, so that the fracture connection can be completed as a smooth path.

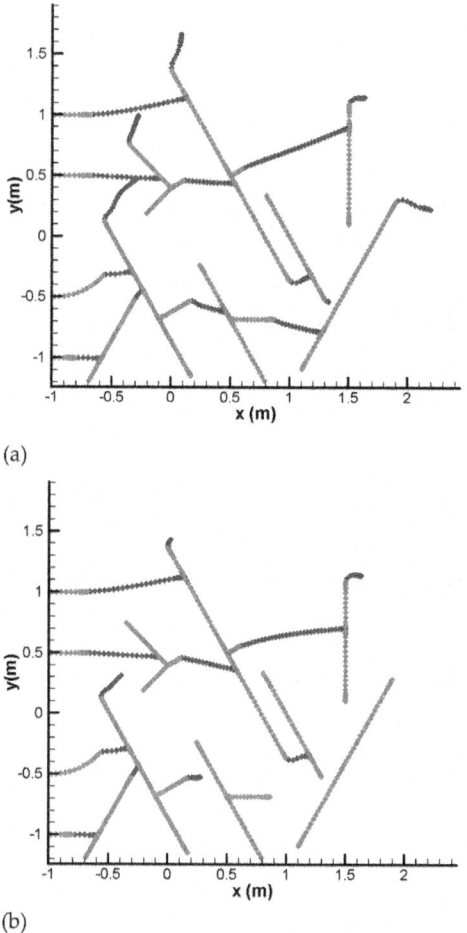

(a)

(b)

Figure 2: Fracture pathways at the time of breakthrough for two different injection conditions: (a) Type I and (b) Type II. The fracture system is subject to the applied stresses $\sigma_{xx}^{\infty}=6$ MPa, $\sigma_{yy}^{\infty}=4$ MPa and $\sigma_{xy}^{\infty}=0$. The initial hydraulic aperture ϖ_0 for all natural fractures is 0.01 mm. The fluid comes into the area from four entry fractures at the left (x=-1 m) and its pressure drives some fractures to propagate and connect the separated fractures as the hydraulic fractures grow toward the right outlet zone (x=2 m). The initial fracture configuration is shown in green and the generated fracture segments are in red.

For the sake of convenience in discussion, the above two injection types are respectively named as type I and type II injection. Figure 2 shows the results at the time of breakthrough when the hydraulic fracture emerges on the right side (x=2 m) of the network of natural fractures, for these two injection types. After fracture reorientation to the direction normal to the minimum principal stess, the fracture development through the network zone is complete. As expected, the type I injection condition results in more fracture growth paths and intersections across the network, while the type II condition results in a localized path, with only one hydraulic fracture continuing to grow past the network on the right side. In Figure 2(b) there are two separate unconnected sets of fractures from top to bottom, both of which connect to two of the initial fractures on the left which are connected to (in this case) the equal pressure fluid source. The influxes into the bottom fracture set decrease in time and more of the total volume injected enters the upper fracture set. The upper fracture set is more conductive. We define the plane $x>2$ m as the exit zone. There is one outlet fracture in this case. However, there are two fluid outlets or extraction sites in Figure 2(a) for type I injection. On the other hand, for the type II injection, the new fracture segments are mainly created along the upper fracture path.

Figure 3 shows the outlet flux variations in time for the two cases used in Figure 2. For type I injection, only around 43 percent of injected fluid has passed through the fracture system to the outlet at the end of simulation period and the other 57 percent is contained in fracture branch inflation or growth in the network. It is also found in Figure 3 that in this case the outlet flux (sum of outlet 1 and outlet 2 for case (a)) increases at a very slow rate at the large time. More fractures under type I injection are connected with each other as a result of crack growth near the fluid source and some fractures continue to grow after the main conductive channels are developed. Fracture segments not opened also store some fluid because of their initial conductivity. The rate that fluid is stored in the fracture network differs between the type I and type II injection conditions. For type II injection, more than 55 percent of injected fluid exits the outlet fracture at the end of the simulation and this outlet flux is increasing at a very large rate as shown in Figure 3 so that the rate of fluid volume stored in the system will become extremely small soon in light of this trend.

In addition, the breakthrough time for these two injection types is provided in this figure. The type II injection has a much earlier breakthrough time so that the fracture growth through the fracture network is more rapid.

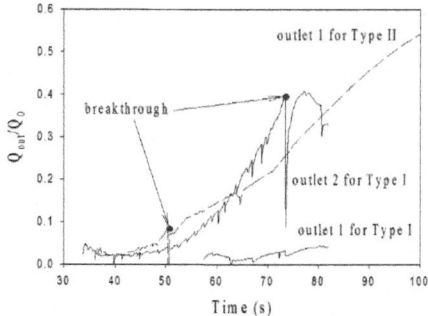

Figure 3: Evolution of normalised outlet flux that is defined by the entrance flow rate of the outlet fractures by the injection rates for two cases provided in Figure 2. There are two outlets for case (a) and they are numbered from the top to the bottom. For case (b) there is only one outlet at the top.

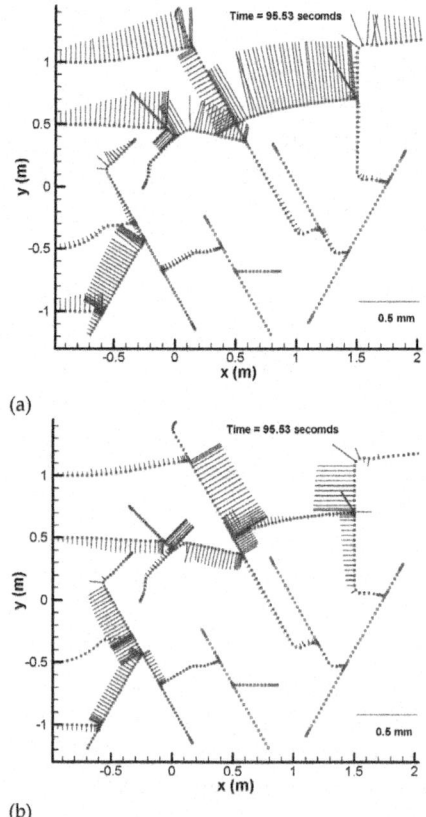

Figure 4: Fracture trajectories and opening (a) and slip (b) profiles, which are represented by the blue bars perpendicular to the fractures, under type II injection at a specific time.

The rapidly increasing trend in outlet flux is also reflected in the opening profiles as shown in Figure 4(a) with the wider open fracture path corresponding to the path that carries the most fluid. The wider fracture channels are localised along the preferential pathway along the upper fracture set to the right-top fracture outlet under type II injection. It is found that at the given time instant, up to 85 percent of the injected fluid is pumped into this preferential path and the fractures not included in this path are all static at late times in the simulation. At the larger times simulated, most fluid just passes through this highly conductive channel to reach the exit zone. Thus, the localised flow channel provides the lowest resistance pathway for fluid flow. Here, we note the contribution of residual hydraulic aperture on the fluid movement and storage. As stated above, each natural fracture has a pre-existing aperture of 0.01 mm in the computations. Thus, some fluid enters and is stored in this pre-existing aperture.

It is found that fractures with wider opening have had their opening enhanced by the large slip along the longest oblique natural fracture, which is oriented at 45 degrees to the x-axis, as shown in Figure 4(b). The kinematic transfer between slip and opening assists the fracture opening.

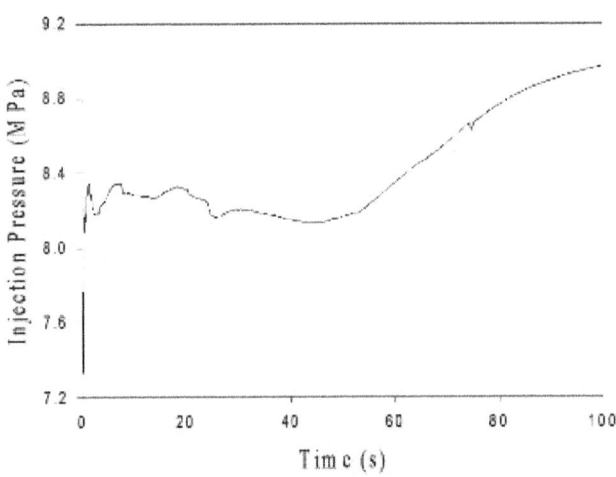

Figure 5: Evolution of injection pressure at the left entry zone for the type II injection method.

Also, it is found that some fracture connections are made even after the preferential fluid pathway is established, since fracture growth can still be generated in the higher pressure region near the entry. After the breakthrough, the injection pressure has to be retained at a higher level, to force fluid through width restrictions that occur at intersections and offsets, as shown in Figure 5. The continuing fracture growth in the network can alter the earlier opening distributions. Comparing Figure 4 and Figure 2(b), it is clear that fracture growth and fracture interconnection can occur inside the network region after the flow breakthrough. Of course, one reason for pressure increasing after breakthrough is attributed to the strong resistance to fluid flow at the last vertical natural fracture on the connected flow path for this specific geometry. This vertical fracture provides the strongest barrier to fluid flow as its opening is highly constrained by the geometry, with the approaching hydraulic fracture making a right angle to the natural fracture. A new fracture would possibly be nucleated at some location along this vertical natural fracture which would reduce this restriction. Although Figure 5 shows the increasing trend of the injection pressure, the pressure will eventually level off or even decrease as the outflux from the network increases to 100 percent of the input rate, as indicated in Figure 3.

DISCUSSION

Fluid-Driven Fracture Nucleation, Growth and Connection

In the model, we only consider hydraulic fracture growth through a finite set of fractures, some of which are initially very small, but potentially provide a conduit with the help of high fluid pressure. The fracture seeds are pre-assumed in this paper and are represented by these small fractures. Therefore, fracture nucleation in a highly stressed area is not dealt with in this paper. This treatment of fracture nucleation can underestimate the fracture number and the fracture connectivity. For the case shown in Figure 4, the lower fluid flow path cannot move to the right outlet and a higher stress level might create new fractures near the entry zone. It would be interesting to study the impact of crack nucleation based on stress conditions rather than just from these pre-existing fractures.

Without considering fluid loss into the rock matrix, hydraulic fracture growth as the main driving force in connecting fractures can create more new segments at the upstream end of the fracture system than at the downstream. In terms of fracture length, both pre-existing and newly created, we can define fracture density as the fracture number per unit area. Although fracture density is a significant measure for fracture connectivity, the longer fractures including the newly created parts would be more important contributors because they are more compliant and will open wider under the same internal fluid pressure [9]. Predicting the early growth of these hydraulic fractures through a pre-exising network, as modeled in this paper, must account for the effect of viscous fluid or incorrect fracture behaviour is predicted. In this model, fracture growth occurs when the failure criterion is satisfied at any fracture tip, with the failure condition defined within the framework of linear elastic fracture mechanics. Fracture curving is the natural result of the local stress field around the tip if the growth follows the maximum tensile stress criterion. Normally, fractures will reorient themselves to the maximum compressive stress direction to increase fracture opening, resulting in local conductivity enhancement as indicated in above results. However, the fracture curving can sometimes lead to intersection of two fractures at a small acute angle, which will make it difficult for the subsequent flow to enter some segments. Sometimes, the subsequently developed sliding on one fracture can seal the fracture channels near the junctions. The development of geometric networks, produced by growing hydraulic fractures, are illustrated by the results obtained above. The results imply that not all connected fractures can contribute to overall conductivity of the system which is contrary to conventional percolation model predictions. These geometric factors affecting fracture growth and fluid flow have been mentioned in early studies [7]. Some fracture growth can occur in the wake of the fracture and flow fronts near the higher pressure entry zone. Local reversed flow has also been observed in the results due to the pressure changes.

Actually, in addition to injection conditions, many other factors such as injection rate and in situ stress can affect the crack growth and coalescence. At the elevated pressure and based on the assumed fracture geometries, one can find that, even through a network of natural fractures, the hydraulic fracture average direction tends to align as much as possible with the direction of the maximum stress. This orientation reduces the viscous dissipation and injection pressure. However, an increase in fluid pressure because of increased rate or viscosity can produce higher pressures upstream of a local offset or restriction which can then lead to opening of cross-cutting natural fractures and branching.

In addition, this paper only considers a limited number of specific initial fracture geometries, the results may be different if the starting geometries are changed and more cases are being considered as a way of making our conclusions stronger and more general. Although a method to deal with network development is presented here, there is a need to work on different geometries to extract some useful general responses for rational simplifications of the expected response for hydraulic fracture network growth.

Implications for Fracture-Controlled Flow System

Our numerical results quantify the overall path of discrete hydraulic fractures growing through a network of pre-existing natural fractures, as shown in Figs. 3 and 4. These results give insight to the behaviour of fracture-controlled flow systems, where the fluid flow and the rock deformation and fracturing are strongly coupled. It is clear that the conductivity depends on the stress-dependent fracture aperture through a strong coupling to fluid flow, as opposed to fixed aperture fractures in conventional percolation models. Local areas can exhibit higher effective permeabilities or strong growth barriers. Such enhanced or restricted opening occur at intersections and offsets, and their existence can affect the total system conductivity, producing a higher pressure level as shown in Figure 5. Early time rapid hydraulic fracture propagation and intersection of small natural fractures establishes a path for the fracture through the natural fracture network, and a single fracture connection event can cause a strong change in the hydraulic fracture channel system that develops.

The model has particular application for understanding the hydraulic fracture connection process through a network. By varying parameters, one finds the transition of fracture-controlled flow pattern from more uniform to more localize and from multi-directional to unidirectional. The hydraulic fractures tend to develop wider and more connective localized fracture channels in establishing a preferred path through the network of natural fractures. In contrast, low rate injection processes that do not involve significant fluid viscous dissipation effects, tend to result in flow occurring along all already connected conductive paths. How to better characterise this difference is still open and to find meaningful parameters in connecting the intricate topological fracture network with diffusion flow patterns requires prediction of propped and unpropped fracture permeability that remains after the hydraulic fracture treatment. The model used here may provide a tool for such parametric studies.

ACKNOWLEDGEMENTS

The authors thank CSIRO for supporting this work and granting permission to publish.

REFERENCES

1.   Berkowitz B., Characterizing flow and transport in fractured geological media: A review. Advances in Water Resources, 2012; 25(8-12), 861–884.

2.   Glass R. J., Nicholl M., J., Rajaram H., Andre B., Development of slender transport pathways in unsaturated fractured rock: Simulation with modified invasion percolation. Geophysical Research Letters, 2004; 31(6), 31–34.

3.   Barnhoorn A., Cox S. F., Robinson D. J., Senden T., Stress- and fluid-driven failure during fracture array growth: Implications for coupled

deformation and fluid flow in the crust. Geology, 2010; 38(9), 779–782.

4.   Dershowitz W. S. Einstein H.H., Characterizing Rock Joint Geometry with Joint System Models. Rock Mechanics and Rock Engineering, 1988; 1, 21-51.

5.   Olson J. E., Joint Pattern Development: Effects of Subcritical Crack Growth and Mechanical Crack Interaction, Journal of Geophysical Research, 1993; 98(B7), 251-265.

6.   Renshaw C.E., Pollard D.D., Numerical simulation of fracture set formation: A fracture mechanics model consistent with experimental observations, Journal of Geophysical Research-Solid Earth, 1994; 99(B5):9359-9372.

7.   Philip Z. G. Jr, Olson J. E., Laubach S. E., Holder J., Modeling Coupled Fracture-Matrix Fluid Flow in Geomechanically Simulated Fracture Networks. SPE Reservior Evaluation and Engineering, 2005; April , 300–309.

8.   Paluszny A., Matthai S. K., Impact of fracture development on the effective permeability of porous rocks as determined by 2-D discrete fracture growth modeling. Journal of Geophysical Research, 2010; 115(B2), 1–18.

9.   Long J. C. S. and Witherspoon P. A., The relation of the degree of interconnection to permeability in fracture networks, Journal of Geophysical Research, 1985; 90(B4), 3087-3098.

10.   Zhang X., Sanderson D.J., Harkness R.M. and Last N.C., Evaluation of the 2-D permeability tensor for fractured rock masses, International Journal of Rock Mechanics and Mining Science & Geomechanics Abstracts, 1996; 33(1), 17-37.

11.   Fu P., Johnson S. M., Carrigan C. R., Simulating Complex Fracture Systems in Geothermal Reservoirs Using an Explicitly Coupled Hydro-Geomechanical Model. 45 th US Rock Mechanics/Geomechanics Symposium, June 26-29, 2011, San Francisco, California.

12.   Zhang X., Jeffrey R. G., Thiercelin, M., Deflection and propagation of fluid-driven fractures at frictional bedding interfaces: A numerical investigation. Journal of Structural Geology, 2007; 29(3), 396–410.

13.   Zhang X., Jeffrey R. G., Thiercelin, M., Mechanics of fluid-driven fracture growth in naturally fractured reservoirs with simple network geometries. Journal of Geophysical Research, 2009; 114(B12), 1–16.

14.   Erdogan F., Sih G., On the crack extension in plates under plane loading and transverse shear, J. Basic Engineering, 1963; 85, 519-525.

15. Pollard D.D., Aydin A.A., Progress in understanding jointing over the past century, Geological Society of America Bulletin, 1988; 100, 1181-1204.

16. Zhang X., Jeffrey R. G., Fluid-driven multiple fracture growth from a permeable bedding plane intersected by an ascending hydraulic fracture, Journal of Geophysical Research, 2012; 117, B12402.

Chapter 2

SUSTAINABLE DRAINAGE SYSTEMS: AN INTEGRATED APPROACH, COMBINING HYDRAULIC ENGINEERING DESIGN, URBAN LAND CONTROL AND RIVER REVITALISATION ASPECTS

Marcelo Gomes Miguez, Aline Pires Veról and Paulo Roberto Ferreira Carneiro

Federal University of Rio de Janeiro Brazil

INTRODUCTION

Floods are natural and seasonal phenomena, which play an important environmental role, but when they take place at the built environments, many losses of different kinds occur. By its side, urban growth is one of the main causes of urban floods aggravation. Changes in land use occupation, with vegetation removal and increasing of impervious rates lead to greater run-off volumes flowing faster. Intense urbanisation is a relatively recent process; however, floods and drainage concerns are related to city development since ancient times. Drainage systems are part of a city infrastructure and they are an important key in urban life. If the drainage system fails, cities become subjected to floods, to possible environmental degradation, to sanitation and health problems and to city services disruption. On the other hand, urban rivers, in different moments of cities development history, have been considered as important sources of water supply, as possible defences for urban areas, as a way of transporting goods, and as a means of waste conveying.

Thus, there is a paradox in the relation between the water and the cities: water is a fundamental element to city life, but urbanisation is not always accompanied by the adequate planning and the necessary infrastructure is generally not provided, leading both to urban spaces and water resources degradation. An interesting historical register illustrates the problem of urban land occupation. In the 16th Century, the architect Giovani Fontana studied the

Tiber River flood in the Christmas of 1598, in Rome (Biswas, 1970). Fontana's conclusions stated that the severe consequences of that flood were related to the occupation of the riparian areas near the confluences of Tiber River with different tributaries and channels, as well as to the lack of information of the people that settled their houses at those places. This situation is pretty similar to what still occurs today: lack of urbanisation planning and control, poor environmental education and the absence of a major framework to unite technical and socio-economic aspects. The main proposition of Fontana to control floods in Rome referred to the enlargement of Tiber River, in order to improve the general flow conditions – a classic view focusing on fast conveying floods to a safe downstream discharge.

As cities started to grow, especially after the Industrial Era, urbanisation problems became greater and urban floods increased in magnitude and frequency. The traditional approach for the drainage systems, which were important as a sanitation measure in the first times of the cities development, conveying stormwaters and wastewaters, turned unsustainable. Flow generation increased and end-of-pipe solutions tended to just transfer problems to downstream. In this context, in the last decades, several approaches were developed, in order to better equate flow patterns in space and time. However, not only the hydraulics aspects are important. Technical measures do not stand alone. The water in the city needs to be considered in an integrated way and sustainable solutions for drainage systems have to account for urban revitalisation and river rehabilitation, better quality of communities' life, participatory processes and institutional arrangements to allow the acceptance, support and continuity of these proposed solutions.

HISTORICAL ASPECTS AND BACKGROUND OF URBAN FLOODS AND DRAINAGE SOLUTIONS

Several ancient civilisations showed great care when constructing urban drainage systems, combining the objectives of collecting rainwater, preventing nuisance flooding, and conveying wastes. During the Roman Empire Age, significant advances were introduced in urban drainage systems. Concerns on urban flooding mitigation and low lands drainage were very important to the city of Rome, which arose among the hills of Lazio region, on the margins of Tiber River. To meet urban drainage needs, a complex network of open channels and underground pipes were constructed. This system was also used to convey people's waste from their living areas (Burian and Edwards, 2002).

During the middle Ages, urban centres suffered a great decay and people tended to live in communities sparsely established in rural areas, near rivers, with minor concerns about urban drainage. Sanitation practices have deteriorated after the decline of the Roman Empire and surface drains and streets were used indiscriminately as the only means of disposal and conveyance of all wastewaters (Chocat et al., 2001). Later, when cities started to grow significantly again, in the Industrial Era, urban drainage found itself regretted to a second plane. The industrial city grew with very few guidelines. The Liberalism influenced urban growth and there was a certain lack of control on the public perspective for city development (Benevolo, 2001). Sanitation, then, became a great problem and inadequate waste disposal led to several sort of diseases and deterioration of public health. The role of urban drainage became very important in helping to solve this problem and, more than often, it was important to fast collect, conduct and dispose securely stormwater and wastewater. Focus was driven to improve conveyance and this was the main goal of urban drainage, until some decades ago. However, considering the fast urban growth of the last two centuries, and the fact that the world population profile is changing from rural to urban, it became hard to simply look at urban drainage and propose channel corrections, rectifications and other similar sort of interventions. Canalisation could not answer for all urban flood problems and, in fact, this isolated action, in a local approach, was responsible for transferring problems more than solving them. The increasing flood problems that the cities were forced to face showed the unsustainability of the traditional urban drainage conception and new solutions started to be researched.

A sustainable approach for drainage systems became an important challenge to be dealt with. Drainage engineers became aware that the existing infrastructure was overloaded. Focus on the consequences of the urbanisation process, that is, the increase of flow generation, which concentrates on storm drains, should be changed. Source control, acting on the causes of flooding and focusing on storage and infiltration measures, emerged as a new option at the end of the 1970s (Andoh & Iwugo, 2002).

An integrated approach, considering the watershed as the planning unit, may be considered the initial basis for a sustainable system design. The design of an urban drainage system integrated with city development, aiming to reduce impacts on the hydrological cycle, acting on infiltration processes and allowing detention on artificial urban reservoirs, joining concerns, restrictions and synergies from Hydraulic Engineering and Urbanism appears as a fundamental option to treat urban floods. Besides the quantity aspects, the water quality became also a main issue and waste waters and solid waste disposal became matters to be treated together. The first flush and the washing of the catchment also introduced a new perception related to the diffuse pollution. At last, and in a complementarily way, rain water harvesting appears as an opportunity to increase water resources availability in urban environments.

Several different conceptions have been proposed in the last decades, with some minor differences among them. All of them, however, tend to consider those questions in an integrated way, trying to rescue natural characteristics of the hydrological cycle, while adding value to the city itself.

Coffman et al. (1998) proposed a design concept of Low Impact Development (LID). LID design adopts a set of procedures that try to understand and reproduce hydrological behaviour prior to urbanisation. In this context, the use of functional landscapes appear as useful elements in the urban mesh, in order to allow the recovering of infiltration and detention characteristics of the natural watershed. It is a change in the traditional design concepts, moving towards a site design that mimics natural watershed hydrological functions, involving volume, discharge, recharge and frequency. The main principles of this approach may be briefly described by the following points:

- minimise runoff, acting on impervious rates reduction and maintaining green areas;

- preserve concentration times of pre-development, by increasing flow paths and surface roughness;

- use of retention reservoir for peak discharge control and improve water quality;

- use of additional detention reservoirs to prevent flooding, if necessary.

In a similar way, another early trend in the drainage system design evolution involved the use of stormwater Best Management Practices (BMP). The term Best Management Practices is frequently used in the USA and Canada and its origin is related with pollution control in the field of industrial wastewaters. Later, it was also referred as a possibility of nonpoint source pollution control and then associated with stormwater management. This way, stormwater BMPs are supposed to work in a distributed way over the watershed, integrating water quantity and water quality control aiming to mitigate effects generated by land use changes, with optimised costs. BMPs are designed to reduce stormwater volume, peak flows, and nonpoint source pollution through infiltration, filtration, biological or chemical processes, retention, and detention. They also may be classified into structural, when referring to installed devices and engineering solutions, or non-structural, when related to procedures changes, like limitations on landscaping practices (US EPA, 2004).

LID and BMP are very often used together and may complement each other. Batista et al. (2005), in Brazil, consolidated the concepts of Compensatory Techniques in urban drainage design, which meant the introduction of several different measures, focusing on infiltration and storage capacity, with the aim of compensating urban impacts on the hydrological cycle.

Another possibility of improving urban drainage solutions concerns the Sustainable Urban Drainage System (SUDS) concept. In this case, the ideals of sustainable development are included in the drainage system design process, that is, impacts on the watershed due to drainage solutions may not be transferred in space or time. Moreover, besides contributing to sustainable development, drainage systems can be developed to improve urban design, managing environmental risks and enhancing built environment. SUDS objectives account both for reducing quantity and quality problems and maximising amenities and biodiversity opportunities, which form the three way concept: quantity – quality – amenity & biodiversity. All of them have to be managed collectively and the desired solution appears in the interface of these three objectives (CIRIA, 2007). The philosophy of SUDS, similar to LID, is also to replicate, as well as possible, the natural conditions of pre-development site.

The key elements for a more sustainable drainage system consider to:

- Manage runoff volumes and rates, reducing the impact of urbanisation on flooding;

- Encourage natural groundwater recharge (where appropriate);

- Protect or enhance water quality;

- Enhance amenity and aesthetic value of developed areas;

- Provide a habitat for wildlife in urban areas, creating opportunities for biodiversity enhancement;

- Meet the environmental and the local community needs.

The continuous evolution of all these concepts and the seek for new urban drainage system solution led also to the Water Sensitive Urban Design (WSUD) concept, initially developed in Australia. Wong (2006) states that the definition of WSUD appears to be confusing among practioners because of its wide range of applications. WSUD tries to integrate social and physical sciences in a holistic management proposition for urban waters.

Langenbach et al. (2008) define WSUD as the "interdisciplinary cooperation of water management, urban design and landscape architecture which considers all parts of the urban water cycle, combines water management function and urban design approaches and facilitates synergies for the ecological, economical, social and cultural sustainability." According to Wong (2006): "WSUD brings 'sensitivity to water' into urban design. The words 'water sensitive' define a new paradigm in integrated urban water cycle management that combines the various disciplines of engineering and environmental sciences associated with the provision of water services, including the protection of aquatic environments in urban areas. Community values and aspirations of urban places necessarily govern urban design decisions and therefore water management practices".

WSUD is centred on integration at a number of levels (ibid):

- The integrated management of potable water, wastewater and storm-water;

- The integration of the urban water management from the individual allotment scale to the regional scale;

- The integration of sustainable urban water management with building architecture and landscaping;

- The integration of structural and non-structural sustainable urban water management initiatives.

Integration of urban water uses in different spatial scales, with the involvement of different knowledge areas, encompassing hydraulic engineering, urbanism, architecture, social sciences and economy, trying to preserve natural environment and adding value to the built environment, in a participative framework where communities play an important role, seems to be the main point to characterise the WSUD concept. The institutional arrangements are key elements here, in order to manage this process.

Furthermore, actions on urban rivers revitalisation or, in a more optimistic sense, actions to allow urban rivers rehabilitation, also arise as a new possibility. The river revitalisation usually includes solutions for the built environment, reconnecting it to the city, but not necessarily recovering natural patterns. The concept of river rehabilitation, however, tries to integrate the river hydrology and morphology, the hydraulic risks associated to the flood control, the quality of waters and the ecological state of the river. These are very complex tasks to be dealt in urban environments, due to several constraints accumulated over time. River canalisation, flood plains disconnection, lack of free spaces, combined sewers (or even uncontrolled wastewaters disposal, as it happens frequently in developing countries), social pressures and other questions appear as difficulties in the way of a river rehabilitation. By the way, one possible decision in urban development may be to state a vision for the river and how to integrate it with the built environment and try to do the best possible to walk in that direction.

Gusmaroli et al. (2011) propose the adoption of an ecosystem approach, in order to supplement or replace the concept of Waterfront Design. The Waterfront Design mainly aims to recover the relationship between river and city around the line where they meet. Stepping ahead brings the opportunity to propose the river rehabilitation concept from the point of view of an environmental improvement, looking at the city as an organism in constant transformation and, therefore, capable of modelling and adapting itself (even only in part) to the demands of recovering more natural features of the watercourses. In this sense, it is a challenge to find ways to recover more natural rivers and rethink the city's growth as a result.

INTERFACE BETWEEN RAIN WATERS AND THE CITY

The urbanisation process changes significantly the natural water balance equilibrium. Vegetation removal and its substitution by impervious surfaces reduces the infiltration possibilities, increasing superficial flow volumes. Besides, natural retention is reduced and the runoff is able to travel faster over regularised urban surfaces. In general terms, even when the urbanisation process is conducted within planning standards, the superficial volumes are greater. If the urbanisation does not account for more sustainable patterns, the peak flow is much greater than the natural one and the peak time occurs early.

Figure 1: Riverbanks occupation in the metropolitan area of Rio de Janeiro City, Brazil (Photo Miguez, 2010).

Uncontrolled urban development, especially in developing countries, where later industrialisation led to a very fast process of city growing, frequently faces the occupation of the natural flood plains and even of the river banks, as it can be illustrated by figure 1. This fact worsens even more the problem of urban floods, because the space needed for flood overbank flows is now occupied by houses, streets and amenities. In this situation, floods tend to spread for larger areas, trying to find room, while affecting urban life in several aspects (sanitation, health, traffic, housing…) and producing great losses. Once the flood space become limited by urbanisation, the flowing waters try to find other paths, inundating areas not subjected to floods before.

Usually, after the first impacts of urban development, changing urban land use and producing the first floods, the drainage canalisation appears as one of the most frequent consequences, both in the allotment level, with the micro-drainage systems, and in the catchment level, with the major works of macro-drainage canalisation. Canalisation works are frequently related to roads and regular grids of the cities. Canalisation, however, as discussed in the previous topic, tends to solve floods locally, with a partial vision of the problem.

In general terms, the occupation process of a watershed normally starts downstream, at the lower and flat areas. The imperviousness effects and the urbanisation towards riverine areas lead to the first canalisation works as a solution for flood control and urban design in these areas, aggravating floods on the basin outfall. With the continuity of the urbanisation process, the upstream areas start to be also occupied, repeating the formulas of the downstream areas. Thus, when these new developments areas start to approximate of the riverine areas and suffer from flooding, new canalisations are settled, and the older downstream areas, where the city centre lies, become flooded again. At the end of this process, the natural storage areas are now occupied, all the catchment is canalised, there are no more flood plains, the channels do not have more discharge capacity, flood is transferred to downstream and large portions of the urban surface are inundated. Besides, the city strangles the drainage system and there are very few possibilities of new canalisation works.

URBAN DRAINAGE DESIGN

Urban drainage design is a relative simple task, when considering the implementation of a project in a new urban area. Channels and pipes are integrated in order to convey the calculated discharge for the design rainfall. Infiltration measures and/or reservoirs may be predicted aiming to keep the generated discharge under a certain limit, and other sort of controls may be imposed, providing a low impact development. In this case, design is made by sub-catchments that are combined and summed in a certain pre-defined order, composing the urban land contribution to the drainage net, in a sequential calculation process. When a drainage system already exists and fails, however, it may become very difficult to propose adequate corrective interventions for the system rehabilitation without considering its systemic behaviour. The combination of superficial generated flows with drainage net flows may be diverse. Waters spilling out of the storm drain system may flow through streets, in an unpredicted way. The streets convey these flows to downstream reaches of the catchment, re-entering the drainage net without control. Sometimes, this superficial flow may even reach a neighbouring catchment, accessing other pipes not yet drowned. Other times, water may be temporarily stored inside lower open areas, like parks or squares, as well as inside buildings, in an undesired way. Urban flood control is a matter of reorganizing flow patterns in space and time (Canholi, 2005). Combination of effects is a difficult question to be assessed in the scale of a catchment and, sometimes, the proposed solutions may not be effective. In this context, mathematical modelling may be an important tool to support the design of integrated urban flood control projects.

Urban Drainage Traditional Design

Traditional practices of urban drainage design are based on canalisation works, in order to adapt the system to the generated and concentrated flows. This approach equates the undesirable consequences of the flooding process, which are the greater and faster discharges produced by the built environment. The urban drainage system comprises two main subsystems: micro-drainage and macrodrainage. The micro-drainage system is essentially defined by the layout of the streets in urban areas, acting in collecting rainfall from urban surfaces. The macro-drainage is intended to receive and provide the final discharge of the surface runoff brought by the micro-drainage net. Macro-drainage corresponds to the main drainage network, consisting of rivers and complementary works, such as artificial canals, storm drains, dikes and other constructed structures.

In general terms, the urban drainage system design comprises the following steps: subdivision of the area into sub-catchments; design of the network integrating urban patterns and natural flows; definition of the design rainfall, considering a certain time of recurrence and a critical time of duration, associated with the concentration time of each sub-catchment considered; step by step calculation of design discharges for each drainage network reach through the Rational Method or another convenient hydrological method; hydraulic design of each drainage network reach. Figure 2 illustrates this approach.

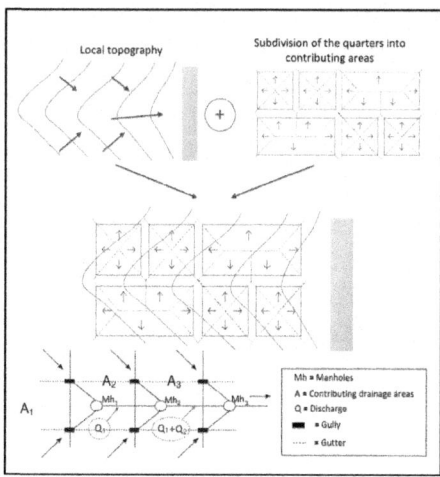

Figure 2: Schematic urban drainage system classic design.

This approach greatly simplifies the real situation and focus only in conveying discharges. Although it may be useful in certain design situations, spatial and temporal effects combination are main factors to be considered when urban floods occur. It is important to have the assistance of a mathematical model as an assessment tool and solutions should be addressed to the catchment as a whole.

Urban Drainage Design Trends

The traditional approach for drainage system design is being supplemented or replaced by newer concepts that seek for systemic solutions, with distributed actions over the catchment, trying to recover flow patterns similar to those that happened prior to urbanisation. Storage and infiltration measures are considered together in integrated layout solutions. Moreover, these new trends add concerns of water quality control, as well as enhance rainwater as a resource to be exploited in an integrated approach for sustainable management of urban stormwaters. Besides, the possibility of combining flood control measures with urban landscape interventions, capable to add value to urban spaces, with multiple functions, is becoming an interesting option from the point of view of revitalising degraded areas, as well as optimising the available resources for public investments.

The vision of integrating urban drainage projects with urban development plans and land use and occupation management, provides a better temporal and spatial range of action for flood control projects, as it seeks to intervene not on the consequence of heavy rains, but on the inundation causes. The changing to a point of view of more sustainable solutions on urban drainage requires a commitment with the future consequences concerning the decisions taken today; so solutions must be flexible enough to allow possible modifications and adaptations in the course of urban development (Canholi, 2005). In urban drainage, sustainability implies that urban floods may not be transferred in space or time. Urban drainage systems have to be planned in an integrated way with urban growth and drainage solutions should be integrated with urban landscape (Miguez et al., 2007). In this context, urbanisation process and urban land use control have both to be thought in order to minimise impacts over the natural hydrological cycle.

This discussion leads to an important point: the understanding on how urbanisation interferes with flow patterns is necessary to develop strategies for stormwater management and urban floods control, by one side, and to establish urban development standards on the other side. Urban drainage planning must consider a broad set of aspects and has to be integrated with land use policy, city planning, building code and all the related legislation. It is possible to say that urban flood control demands the adoption of a varied set of different measures and concepts. Among these measures it is possible to distinguish two greater groups of possible interventions: the structural measures and the non-structural measures.

Structural measures introduce physical modifications on the drainage net and over urban catchment landscapes, like canalisation, dams, reservoirs, urban flood parks, dykes, among others, intending to change the relations between rainfall and runoff and to reorganise flow patterns. Non-structural measures work with environmental education, flood mapping, urbanisation and drainage planning for lower development impacts, warning systems, flood proofing constructions, and other actions intending to allow a more harmonic coexistence with floods. Structural measures are fundamental when flood problems are installed, in order to bring the situation back to a controlled one. Non-structural measures are always important, but are of greater relevance when planning future scenarios, in order to obtain better results, with minor costs.

Under this new perspective, the urban drainage projects, in theory and whenever possible, should neutralize the effects of the urbanisation, restoring the hydrological conditions of the pre-urbanisation, bringing benefits to the quality of life of the population and aiming the environmental preservation.

Structural Measures

Structural measures can be classified according to their performance in the catchment. According to Tucci (1995), they can be divided in distributed measures, measures in the micro drainage and measures in the macro drainage, as detailed below and exemplified in Figure 3.

- Distributed Measures: these measures act on the lot, squares and sidewalks. They are also known as source control measures.

- Measures in the micro drainage: these measures act on the resulting hydrograph from one or more lots.

- Measures in the macro drainage: these measures act on the rivers and channels.

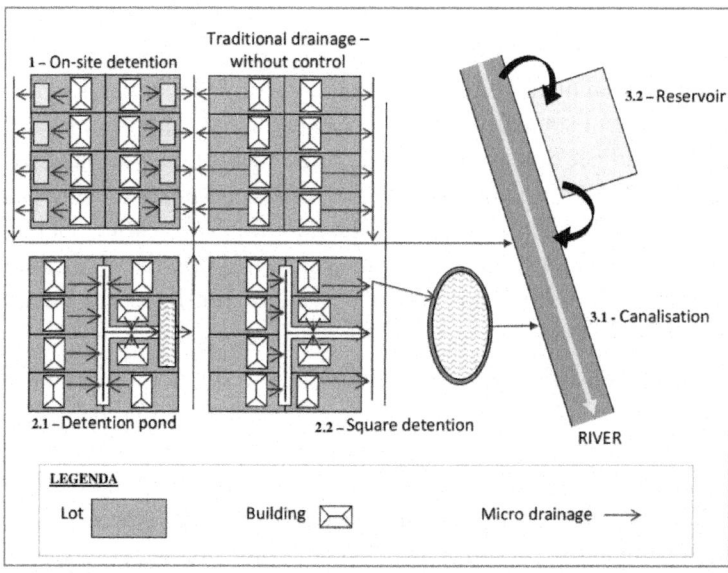

Figure 3: Examples of flood control measures according to their working principle: 1. Distributed Measures; 2. Measures in the micro drainage; 3. Measures in the macro drainage (Rezende, 2010).

Canalisation is, undoubtedly, the more traditional measure adopted in flood control interventions. Its main objective is to improve the hydraulic discharge capacity of the macro drainage network, through the removal of obstructions to the flow on the main channels, the river channel rectification and the revetment of the riverbanks. Another traditional measure widely used to contain river overflow is the implementation of dykes associated with polders, especially in low areas of the catchment, which allow the protection of the urbanised plains. The protected areas, which remain unable to drain the precipitated water over its own local catchment during the river flood events, are generally linked to the main water body by one-way gates (FLAP gates) or by pumping stations.

Thus, it is necessary to preserve unoccupied areas inside the polder to receive and temporarily store these waters.

Another set of measures, as an alternative to the simple improving conveyance, proposes to act with the possibilities of storage and infiltration. Important examples of this set of measures are the detention ponds. These measures used to be generally designed at the upstream reaches of the most urbanised regions, where the occupation still is sparser and where there are available areas for the implementation of the ponds. In situations where the urbanisation occupies every available space, the detention ponds have been adapted to other scales, allowing the use of public spaces such as parks, parking lots and squares, in order to temporarily store the rainwater from less frequent events and also add value to the urban environment and region, as can be seen in Figure 4, showing a detention pond implemented in Santiago (Chile), which is associated with a landscaping design. The use of this kind of structure has a very wide spectrum, and it may be used through the implementation of large ponds, or by the distribution over the watershed of several of these devices, where they can act in squares or even inside the lots.

Figure 4: A detention pond in Santiago, Chile (Photo Miguez, 2009).

Another possibility for stormwater storage measures may be the use of reservoirs that have the goal of improving water quality. These structures are the retention ponds and the constructed wetlands. It should be noted that the main objective of these measures is the treatment of rainwater, remaining smaller its quantitative effects. This is due, in part, by the need to provide a permanent pool and also a greater time of permanence of the water inside the reservoirs, to enable the treatment processes with the required efficiency. Figure 5 presents a picture of a retention pond, constructed in the city of Lagord, France. This pond is part of a drainage plan of the whole region of La Rochelle, which includes the city of Lagord, aiming for the treatment of rainwater.

Figure 5: A retention pond in Lagord, France (Photo Rezende, 2009).

Measures that aim to favour infiltration processes of the rainwater in the ground, allowing the partial recovery of the natural hydrology of the watershed, are also interesting options in the context of flood control. They may assume different configurations, according to its operation. An important measure in this context, also because its environmental implications, may be the reforestation of degraded areas, such as hill slopes and riverbanks that have been illegally occupied. Miguez & Magalhães (2010) indicate that reforestation prevents soil erosion, preserves the superficial soil layer and promotes the infiltration and, thus, the volume of runoff is reduced, allowing the correct functioning of drainage structures, since smaller amounts of water and sediments reach the system.

Non-Structural Measures

Structural interventions for flood control do not provide a complete risk protection for the design areas. These areas may be still subjected to flood events with a magnitude greater than that of the protection designed level. So, measures that aim to prevent the population from these risks and help them to deal with flooding are necessary. Unlike the structural measures, which act physically changing flow relations, the non-structural measures goal is to reduce the exposure of lives and properties to flooding. A wide range of possible actions, from urban planning and flooding zoning until individual flood-proof constructions, compose this set of measures.

Johnson (1978) identified the following non-structural measures: installation of temporary or permanent sealing in the opening parts of the buildings, elevation of pre-existing structures, construction of new elevated structures, construction of small walls or dykes surrounding the structure, relocation or protection of goods that could be damaged within the existing structure, relocation of structures out of the flooding area, use of water resistant material in new structures, regulation of the occupation of the flooding areas, control of new community settlements, regulation of parcelling and building codes, purchase of flood hazard areas, flood insurance, implementation of forecasting and flood warning systems with evacuation plan, adoption of tax incentives for a prudent use of the flood area, and installation of alerts in the area.

SUSTAINABLE URBAN DRAINAGE SYSTEMS CONCEPTS

The urban drainage system needs to be viewed in an integrated way in the context of the sustainable urban development. It is crucial to understand the crossed relationships between urban growing and flood problems. The aspects involved vary from environmental conservation, land use control, low impact development, and healthy city life. To achieve these goals, related to a sustainable urban drainage, however, it is necessary to construct a framework integrating legal, institutional, social, technical and economic aspects. In this context, it is important to clearly identify the applied regulation in terms of urban zoning and land use control, the water resources policies and the water resources management practices, the integrated environmental sanitation opportunities and constraints, the building standards and limitations, the role of the institutional agents and community participation. Areas to be protected from urban growth need to be delimited, as well as, sometimes, it will be important to recover areas already occupied. This is not simple, due to social pressures against possible dwelling relocation procedures. In developing countries, for example, it is common to have "informal" cities, conforming slum areas growing on risky situation, along riverine areas or hill slopes. Landscape changing should be minimised and original river characteristics could be recovered. In this context, preventing urban land to be heavily impervious is one of the major goals. Minimising impacts on the urban water cycle is of fundamental importance. The development project must comply with natural hydrological aspects or provide compensatory measures for urbanisation changes. The watershed, as a complex and integrated system has to be considered as a whole, not only in physical terms. This must be the unit of planning and design. The documents that integrate urban development and the strategies for a sustainable drainage system, and, therefore, a sustainable city, are the Urban Development Master Plan and The Urban Stormwaters Master Plan

The non-existence or non-fulfilment of plans for urban development leads to drainage and flood control projects that are restricted to emergency, reactive, and sporadic actions, defined just after the occurrence of disasters (Pompêo, 2000). This scenario, usually based on the simple and quick removal of water from highly impervious areas by means of canalisations, has become unsustainable and requires a new vision into the problem of urban flooding. Holz & Tassi (2007) suggested that the current drainage system should dismiss the solution of simple removal, as fast as possible, of the non-infiltrated stormwaters that come from the increasing in soil imperviousness, replacing it with measures aiming to mitigate the impacts of this process by facilitating the infiltration and water retention in order to regenerate the hydrological conditions of pre-urbanisation. They emphasize, however, the importance of combining the use of traditional and unconventional drainage structures in order to optimise the system. The fall of the old paradigm should not be simply substituted by a new one. The traditional techniques have to be adapted to a new use, adding the accumulated knowledge to the sustainable solution. Pompêo (2000) emphasizes the need to think the activities related to mitigation of floods in a preventively way, highlighting the value of planning applied to flood control projects. In this context it is introduced the ecosystem approach, which represents the evolution of the reactive thought of the conventional Drainage Master Plan for a proactive and advanced thought in the form of management of the natural and built environment, considering them as interdependent and integrated components. An ecosystem approach can result in lower costs, since it seeks to reduce the need for costly and complex actions of remediation, emphasizing the orientation and planning decisions on land use changes. This option tends also to reduce costs of maintenance over time, because more natural arrangements tend to work by themselves.

This new vision has been based on the concept of Sustainable Development. It's possible to affirm that these systems are designed to both manage the environmental risks of urban stormwater and contribute, whenever possible, to an improved environment and quality of urban life. Sustainable drainage projects aim to reduce runoff through rainwater control structures in small units. This way, the runoff control performed on source reduces the need for large structures of mitigation and control on the river channels.

Environmental issues presented today to the cities highlight the failure of technical solutions for urban drainage projects, demanding a new approach which should focus on the problem of urban flooding by incorporating the social dynamics and the multisectoral planning in the search of solutions (Pompêo, 1999).

The urban infrastructure systems are interdependent and the fact of not considering the effects of one system over another, or of a system over the urban environment, can reduce the efficiency of these systems or even turn not viable their operation, as is the case of the relations between Drainage and Flood Control systems and Water and Sanitation, Solid Waste, Land Use and Housing systems. Two trivial examples in peripheral countries are the failure of the drainage system operation by the inefficiency or lack of management of municipal solid waste; or the deterioration of health by the inefficient sanitation conditions which, combined with flood events, provide the proliferation of water related diseases.

Problems associated with urban drainage systems are not exclusively technical, but primarily of institutional nature, such as the lack of cooperation between the different departments responsible for urban management and the communication between the city and its citizens (Stahre, 2005). The lack of cooperation between departments may arise from conflicting interests and priorities.

According to Stahre (2005), the Drainage Department of Malmo, Sweden, sees the approach of the sustainable drainage system as an ideal way to achieve its goals and objectives with a low cost, if compared to the traditional drainage system, whereas the Department of Parks and Recreation sees the sustainable drainage solutions as a good ally for the development and improvement of quality of life in the urban environment by increasing the value of urban parks. As a consequence of this sharing of interests, the costs of structures, deployment and maintenance can be divided between the two departments, and also with others that will share the benefits of implementing this system. Stahre (2005) concluded that the solution of rainstorm problems can no longer be regarded as a simple technical service supported only by the Drainage Department, because these waters now represent an important positive resource to the population, inserted in the urban environment. It's important to emphasize the value of water in the city as a demanded resource, not wasting its potential uses.

The failure to consider the drainage in the urban development plans may result in more expensive solutions for flood control, often noneconomical. The successful implementation of a Sustainable Drainage System depends on the cooperation among the different technical departments responsible for urban planning and the active participation of the population. In this context, the existence of compatibility among the Urban Development Sanitation, Solid Waste and Urban Drainage Master Plans, aiming the integrated planning of the city should be ensured.

The design of the Urban Stormwater Management currently presents, according to Righetto et al. (2009), the aggregation of a structural and non-structural set of actions and solutions, involving large and small works and planning and management of the urban space. The Stormwater Management Plan of the City must necessarily meet the principles of Sustainable Management of Urban Stormwater, and should seek the following objectives (Ministério das Cidades, 2004):

- Reduce the damage caused by floods.

- Improve the health of the population and of the urban environment, within the economic, social and environmental principles.

- Plan the urban management mechanisms for the stormwaters and the municipality river network sustainable management.

- Plan the distribution of stormwaters in time and space, based on the trend of evolution of the urban occupation.

- Regulate the occupation of areas at risk of flooding.

- Partially restore the natural hydrological cycle, reducing or mitigating the impacts of urbanisation.

- Format an investment program of short, medium and long term.

The drainage projects proposed by the plan should provide the most cost-effective relation, covering social and economic aspects, as well as being integrated into the guidelines of the local River Basin Committee, if any. The plan must also contemplate a socio-environmental work, through the development of a project that addresses social mobilization, communication, training of educators / agents in the area of environmental sanitation and other environmental education activities, aiming a social-economic and environmental sustainability, including the community participation in the phases of design, implementation, evaluation and use of the proposed works and services.

The premises to be considered in the creation of the Stormwater Management Plan are (Ministério das Cidades, 2004):

- Interdisciplinary approach in diagnosing and solving the problems of flooding.

- Stormwater Plan is a component of the Urban Master Plan. As drainage is part of the urban infrastructure, it should, therefore, be planned in an integrated way.

- Runoff cannot be intensified by the occupation of the basin.

- The Plan has as its planning unit each watershed of the city.

- The stormwater system should be integrated into the environmental sanitation system, with proposals for the control of solid waste and the reduction of stormwater pollution.

- The Plan shall regulate the territory occupation by controlling the expansion areas and limiting the densification of the occupied areas.

- This regulation should be done for each watershed as a whole.

- Flood control is a permanent process and should not be limited to regulation, legislation and construction of protection works. A plan to monitor and maintain the proposed measures is needed along time.

It is important to schedule the actions of the Stormwater Management Plan in time, recognizing short, medium and long term actions, to ensure enduring solutions for drainage. Short-term measures intend to correct or mitigate the immediate problems of macro-drainage network, promoting the removal of singularities, desilting and maintenance of the original characteristics of the system. Measures to prepare the required database for the consubstantiation of the plan are also important. The construction of a mathematical model should also be among the initial activities, to provide a systematic evaluation tool. Also among the short-term activities, it is necessary to make a diagnosis for the watershed.

After that initial stage, in the short/medium term, measures should be designed to control runoff at the source and for the recovery of the natural hydrological cycle characteristics. These measures should be scheduled, to be implemented over time. Flood maps should be provided for the evaluation of the proposed scenarios and, also, allow the definition of interactions with land use zoning. From the developed experiences, it may be possible to produce a practical drainage manual, in the medium/long term, bringing together recommendations for all the developments and future projects. The long-term actions should sustain an adequate operation for the drainage system, through maintenance and monitoring. Environmental education campaigns and community engagement are also needed to help in supporting the proposed solutions.

River Revitalisation

It is known that self-sustaining river systems provide important ecological and social goods and services to human life (Postel and Richter, 2003 apud Palmer et al., 2005). River revitalisation is an issue that comes as a necessity to face the progressive deterioration of river ecosystems worldwide. The results can be analysed not only from the aesthetic viewpoint and to improve the environment, but also in terms of hydrologic and ecologic functioning of restored river reaches, increasing the quantity and quality of river resources and their potential use to riverine population (González del Tánago & García de Jalón, 2007). In urban areas, River Revitalisation is more complex, because of the large modifications suffered by the riverine areas, with the construction of buildings and roads, which make it difficult to have the space needed to recover the natural processes of the river bed and its banks (ibid). The river revitalisation process needs to be discussed in a particular way for urban areas and a consensus solution between the natural landscape and the built environment must be found.

In highly urbanised regions, generally the available areas for interventions are scarce, there are numerous socio-economic problems, which make the revitalisation process very complex, because it involves the need of large riparian areas, in order to find space for the river recover its natural course and flooding areas. However, even if these riparian areas were restored to the original natural condition, the heavy modifications that the catchment suffered over the time would probably lead to floods still happening. Thus, the space that would be required to recover the river course functions today is greater than the natural one. Actions in the basin have to be considered, to decrease the imperviousness and to rescue superficial retentions, with the use of reservoirs.

Urban water courses restoration is a challenge for managers, researchers, experts and citizens. In urban environments, the main focus must be on the restoration of the lateral connectivity with the river banks and its tributaries, the restoration of the river natural flow regime, as well as the increase of the degree of freedom of the river. The combination of flood risk management concepts with River Revitalisation measures can be a solution of efficient applicability in urban rivers, in comparison with traditional and localised drainage solutions (Jormola, 2008). Thus, what is expected is the creation of a self-sustainable natural river system, in order to maintain the flood control function after the flow patterns restoration. In this context, a sustainable approach for drainage system should consider River Revitalisation as one of the tools aligned with this major objective.

However, it's important to note that, even when the adopted measures configure only a partial revitalisation, they are important. In addition to reducing the peak flood, they help in the dissemination of this kind of techniques and provide a new perception about the existence of the river for the involved community. They also allow the river valuing and reintegration as part of urban landscape. Finally, it should be pointed out that any process of revitalisation takes time to be fully developed and find it complete. It is necessary to await the responses of the environment, regarding the "new" conditions to which it was submitted. During this time, complementary actions, resulting from the monitoring of this evolution may and should be developed.

INSTITUTIONAL ASPECTS SUPPORTING THE INTEGRA-TION OF A SUSTAINABLE URBAN DRAINAGE SYSTEM AND THE WATER RESOURCES MANAGEMENT – THE BRAZILIAN EXPERIENCES MANAGEMENT

The legal framework comprising the different institutional and management levels is probably the first arrangement to be settled in the path of the sustainability. One of the questions that first arise refers to the urbanisation responsibilities. City development is an attribution of the Municipality, while water resources management is something that needs to consider the basin scale and, generally, comprises a regional planning. In Brazil, legislation shows a significant concern in providing tools for guiding the different institutional levels in the path to build a sustainable city. However, Brazilian cities, in general, do not present an adequate quality level for the built environment and the city life. In order to have a general view of the Brazilian legal framework for urban development and, in particular, for the achievement of sustainable urban drainage systems, a brief review is included as an introduction to this section, in the following lines.

The Federal Urban Land Parcelling Act (Brazil, 1979) establishes the minimum standards for urban developing. This Act considers that a plot is a parcel of urban land provided with the basic infrastructure, meeting the urban restrictions. This basic infrastructure refers to urban drainage, sewerage, water supply, electricity and public roads. The Act also states that it is not allowed to have allotments in flooding areas, in environmental protected areas or in polluted degraded areas. Another Federal Act, known as "City Statute" (Brazil, 2001), establishes detailed rules for public land use order and social interest that regulate the use of urban property in favour of the collectivity, the security and well-being of citizens and environmental balance. The urban policy aims to organise the fulfilment of the social functions of the city and of urban property by the application of a set of general guidelines, from which the following topics are detached for the purposes of the present discussion:

- guarantee the right to sustainable cities, meaning the right to urban land, housing, environmental sanitation, urban infrastructure, transport and public services, work and leisure for present and future generations;

- democratic management through people's participation and associations representing various segments of the community in the formulation, implementation and monitoring of plans, programs and projects for urban development;

- planning the development of cities to prevent and correct the distortions of urban growth and its negative effects on the environment;

- supply of urban infrastructure and community equipments, transport and public services to serve the interests and needs of the population;

- ordering and control of land use to avoid pollution, environmental degradation and excessive or inadequate use in relation to urban infrastructure;

- protection, preservation and restoration of the natural and built environment, and cultural, historical, artistic and landscape heritages.

Several important urban management tools are made available in the context of the City Statute. The Urban Master Plan is considered to be the basic instrument for the urban developing policy.

In the field of water resources management the "Water Act" (Brazil, 1997) defines that the hydrographical basin is the unit of planning and design for water resources purposes. Another important reference is the "Basic Sanitation Act" (Brazil, 2007). This Act establishes the first national guidelines for the federal policy of basic sanitation. An important key element of this Act is the integrated conception of the sanitation services and their relation with the efficient management of water resources. For all purposes this Act considers sanitation as a set of services, infrastructure and operational facilities for drinking water supply; sewerage collection, treatment and disposal; urban solid waste management; urban drainage and storm water management.

As it was said in the beginning of this section, although legislation has merits, there are difficulties in urban development and control. This discussion and the possible ways to go further in urban sustainability, concerning urban waters, are the main points of the next topics.

THE LOCAL PLANNING AND THE MANAGEMENT LEVEL

The jurisdiction of the municipality in federative countries focuses on roles that are generally related to the provision of local public services and to planning, supervision and development functions, which are related, among others, to land use planning, environmental protection and also to a certain level of economic activities regulation (Dourojeanni and Jouravlev, 1999). Considering the Brazilian case, recently the municipalities with larger investment capacity began to incorporate roles related to the provisions of social services that used to be traditionally restricted to the state and federal levels. It is observed from the 1990's a tendency to extend the role of local public levels regarding the environment management. A lot of factors, however, limit the performance of the municipality in the water management. In Brazil, for instance, there are legal constraints determined by the Federal Constitution, where the cities cannot directly manage the water resources contained within their territories, except for certain formal agreements that transfer some assignments through cooperation arrangements with the State or the Union. Water resources management is a matter of river basin management and cannot be restrained to an administrative territory jurisdiction. Thus the role of the cities is restricted to lower levels of relevance and administrative autonomy (Jouravlev, 2003). Municipality participation in basin organisms for water resources management (called Basin Committees, in Brazil) has been the main stage where the interactions between Municipalities and the other public or private actors occur.

Despite of the fact that the administrative level of the municipality is the closest to social reality, its range of political and administrative roles does not allow a systemic vision of the territory in which it is inserted. In turn, the absence of a clear definition of the extent of local governments functions, in general, linked to the traditional tasks of territorial administration, the supervision and provision of local services, and the fact that most of the municipalities have a reduced financial autonomy, depending on transfers from other government levels, makes it difficult to them to have a more effective participation in the water management. Referring to the financial constraints, Lowbeer & Cornejo (2002) warn that the multilateral financing agencies, except for the Global Environment Facility -GEF, have not yet come to explicit in their agenda the need for integrated management projects of natural resources linked to the territory management and land use, particularly in urban areas. Few are the experiences implemented which have coordination between water conservation/preservation and regulation of the land use against the (dis)functions of urban growth.

Another aspect is that the essentially local nature of the city governments' interests makes them act more like water resources users rather than managers of these resources (Jouravlev, 2003). These aspects are exacerbated in metropolitan areas where local governments have often antagonistic interests and priorities, creating dissent environments with little room for cooperation. The metropolitan question is surely one of the greatest present challenges.

Even if there are restrictions to the municipalities' participation as direct managers of water resources, there is no doubt about the importance of local governments on planning and ordering the territory, due to its consequences on the water resources conservation. It is the county attribution the elaboration, approval and supervision of instruments related to land use zoning and planning for development purposes, such as the Urban Master Plan, the delineation of industrial, urban, rural and environmental preservation areas inside the municipality, land parcelling criteria, the development of housing programs, among other activities that have impact on water resources and sanitation conditions, especially in predominantly urban watersheds.

According to Peixoto (2006), the history of the production process of the urban space and its impact on natural resources and human dwellings quality demonstrates the difficulties for articulation between urban and environmental issues. At the same time, however, it may be observed a trend of convergence of these issues towards sustainability, expressed on the federal Acts previous described in the beginning of this topic. Nevertheless, what is observed in the country is the disconnection between the practical instruments of water management and of land use planning, reflecting, perhaps, the lack of legitimacy in the planning process of Brazilian cities as well as certain gaps in local legislation. Several cities are marked by a high level of informality and even illegality in land use – social problems arise as a critical element and urban mesh degrades with spreading slums, also favouring several environmental impacts. According to Tucci (2004), the greatest difficulty for the implementation of integrated planning stems from the limited institutional capacity of the municipalities to address complex and intersectoral issues. However, it is relevant to point out that there are differences among cities, depending on its size, geographic position, and distance from the metropolitan areas, historical urban structuring and evolution, qualification of the technical public staff. Peripheral districts in metropolitan areas, presents, sometimes, an outdated legislation, aggravated by the absence of reliable information and lack of quality technical support.

Integrated Water Resources Management: Interfaces with Sectoral Policies and Territory Planning

The institutional organisation of water resources in Brazil began in the 1930s with the establishment of the Water Code, in 1934. The Water Code represented a milestone in the institutionalisation of the water planning in the country, allowing the expansion of the electricity sector. The granting of hydroelectric developments and electricity distribution services went to the Union with the establishment of this Act. In the same year it was also created the National Department of Mineral Production (DNPN, in Portuguese), within the Ministry of Agriculture, which incorporated the Service of Geology and Mineralogy and the Water Service. During the year of 1938, it was created the National Council of Water and Power, attached to the Presidency that, together with the DNPN, became in charge to decide on water and electrical energy in the country.

Even before the 30's, several governmental committees had been set up in order to coordinate and implement water works. However, the onset of a coordinated action in the water sector has only occurred in 1933, with the creation of the Sanitation Committee of Baixada Fluminense, in the framework of National Department of Ports and Navigation. Baixada Fluminense is an important region, in the metropolitan area of Rio de Janeiro City, characterised by extensive lowlands. This Committee was responsible for the formulation of a comprehensive drainage program for Baixada Fluminense, which was an unprecedented action in the country, with the main aim to make this vast lowland plains of the State of Rio de Janeiro arable and, secondarily, to eradicate yellow fever and to control the local floods (Carneiro, 2003).

This Committee was the seed of the National Department of Sanitation Works – DNOS (in Portuguese), in 1940, created with the responsibility for implementing the national policy of general sanitation, both in rural and urban spaces, including flooding mitigation, erosion and water pollution control, and the recovery of areas for agricultural or industries uses, as well as the settlement of water supply and sewage systems. Despite the range of assignments given to DNOS, its performance was limited, in its initial phase, to the drainage work for drying wetlands, consolidating and expanding the program prepared by the Sanitation Committee of Baixada Fluminense (Ibid).

Dourojeanni and Jouravlev (2001), referring to the experiences of integrated management in Latin America, point out that many of the institutions set up from the 1940's were increasingly incorporating multiple uses of water, even though they used to have as initial particular goals - like flood control, hydroelectric plants, irrigation projects and water supply. Few were those which began their activities by integrating these multiple uses of water.

An important experience in water management in the country, considering the perspective of integrated water resources management, was the creation of the Special Committee for Integrated Watershed Studies (CEEIBH, in Portuguese), in 1978. Supported by CEEIBH, several watersheds committees were constituted. However, despite the important role of these committees on the preparation of studies and investment plans for the recovery and management of the related watersheds, the efforts and experiences have not been able to establish an integrated management of the water resources, nor the implementation of proposed actions could reverse the basins degradation. They also failed to avoid sectoral and fragmented management practices. In part, the low effectiveness of these initiatives was due to the fact that these committees had a merely advisory feature.

Until early 1985, the National Department of Water and Electrical Energy (DNAEE, in Portuguese) was responsible for the management of water resources in the country. Since 1988, however, with the new Federal Constitution, a set of modifications in the water sector was introduced: the definition of Federal/State dominion of the water bodies, the definition of the water as a public good endowed with economic value and the need for the integration of water resources management with land use management policies. In 1995 it was created the Water Resources Secretariat, linked to the Ministry of Environment, with the objective to act in the planning and control of the actions related to water resources in the Federal Government, among others. This institutional change represented the incorporation of the concept of multiple uses of water in the environmental context (CEPAL, 1999).

With the approval of Act 9,433, in 1997, the country received one of the most complete regulatory frameworks focused on water resources managing in the international scenario. The National Water Resources Management System aims to coordinate the administration of water resources in the country seeking to integrate it with other sectors of the economy; administratively arbitrate conflicts related to water uses; implement the National Water Resources Policy; plan, regulate and control the use, preservation and restoration of water resources; and charge for water use, among others. This Act establishes that the watershed is the territorial unit for the National Water Resources Policy implementation and for the National Water Resources Management System actions.

The principles adopted by the Water Act, as the Act 9.433 became known, are adherent to the statements of the major international conferences that have dealt with the water issues and which substantially contributed to the concept of development on a sustainable basis. However, as several authors emphasize (Dourojeanni & Jouravelev, 2001; Cepal, 1999), the integrated management of water resources requires a change on planning paradigms, in both public and private levels. Integrating these variables implies on operate in various fields of public policies, especially in those related to regional and urban development and the institutional arrangements that shape those policies. According to Silva & Porto (2003), the institutional planning and management of water resources system faces four types of integration challenges:

- Integration among activities directly related to the water use in the basin: water supply, wastewater depuration, flood control, irrigation, industrial use, energy production, in order to optimise the multiple use under the perspective of a joint management of water quality and quantity;

- Regulatory articulation with sectoral systems that do not use directly the water resources, such as housing and urban transportation, in order to prevent excess of imperviousness or urban pollution impacting water sources;

- Territorial integration with the instances of urban planning and management, in order to apply preventive measures in relation to the urbanisation process, avoiding the increasing demand of quantity and/or quality of existing water resources, including flood occurrences;

- Articulation with neighbouring basins, celebrating stable agreements on the current and future conditions of imported and exported flows of the waters used in the basin.

New Institutional Arrangements: Watersheds and Metropolitan Areas

The current approach for the water resources management in urban areas presupposes an inseparable and integrated planning for urban development projects. Tucci (2004) proposed an approach where aspects related to watershed protection, sewerage collection and treatment, solid waste collection and disposal, urban drainage, river floods and land use are treated in an integrated way, considering the Urban Master Plan as the central point.

Gouvea (2005) stated that the dynamics of the urban growth, often disordered and even chaotic, was gradually showing the ineffectiveness of many programs and projects implemented in isolated modules and developed from the mistaken idea that the urban reality could be divided and treated in a compartmentalised way. The author notes that actually the city must be seen not only as a specific and complex system, but also as part of a larger system, regional or even national, made up of several subsystems, such as habitation, public transportation, sanitation, natural environment, etc., which are closely related and require an integrated and multidisciplinary approach.

Tucci (2004) lists some factors that hinder the application of the integrated management concepts in cities, as follows:

- Absence of adequate knowledge on the subject: the population and professionals from different fields and levels who do not have adequate information on the problems and their real causes. The decisions usually result in high costs. For example, the use of canalisation works as drainage solutions is a widespread practice in Brazil, even when they represent high costs and impacts. Generally, the channels transfer flood downstream, affecting another part of the community. These works can reach an order of magnitude of 10 times the costs of on-source control measures;

- Inadequate design for urban systems control: an important part of the technicians who work in urban issues is outdated about the environmental concerns;

- Fragmented vision of the urban planning: urban planning and development do not always incorporate integrated aspects related to water supply, sewage, solid waste, flooding and urban drainage;

- Lack of management capacity: the cities are not designed for proper management of the different water aspects in urban areas.

The situation is even more critical in the metropolitan areas that have a high level of conurbation. It is no coincidence that new institutional arrangements for the cities management have aroused the interest of technicians and researchers who identify the need for the resumption of planning on a regional basis, without neglecting, of course, issues that could and should be treated locally. Therefore, the challenges related to urban waters management combined with the intense process of territory occupation, develop into specific problems of the built environment that require a tailored approach. The following sub-items discuss the new institutional arrangements and the perspectives they bring to fill the institutional gap left by the absence of metropolitan instances for intensely urbanised cities planning in Brazil.

Watershed Committees

The central figure in the water resources management system is the watershed committee. The committees are public political organisms of decision making, with legislative, deliberative and advisory powers, concerning the use, protection and restoration of water resources, involving a wide representation of organised sectors of civil society, governments and water users. The committees work as a decentralised locus for discussion on the water uses issues of a watershed, acting as a mediating instance among different interests. These committees are seen as "water parliaments", playing the role of the decision maker within the basin context. The committees composition, as provided by Act 9,433, comprises the Union, the States and the Cities located, even if partially, in the respective basin; the water users within basin area; and the civil water resources entities with activities in the basin.

Nevertheless, it is a fact that the committees established in the country have found great difficulties in fulfilling their decisions and in executing their investment plans. Two main aspects can be identified as constraints to the action of the committees. The first one is that the revenues from charging water uses, which is the only funding source of the committees, are not enough to make the necessary investments for the watersheds recovery. Thus, the committees remain dependent on the traditional sources of investment, which have their own mechanisms for eligibility and prioritisation. The second aspect is that the committees have not gained the necessary political and institutional legitimacy for the public policies coordination related to the watershed, nor could it guide the investments to target actions of its interest. This last aspect stems from the fact that the basin does not constitute a political space reference for Brazilian institutions.

Without disregarding the committee's importance in the public policies decentralisation and in the society participation, the above pointed aspects restrict the possibilities of the committees working as integrators of public regional policies.

Public Consortia

The possibility of forming consortia in Brazil dates from the late nineteenth century; however, there were, over time, numerous configurations and autonomy design of these instances of inter-municipal cooperation. Table 1 summarises the forms of consortia planned in Brazil for over a century.

Table 1: Consortia Models provided in Brazil in the period 1891-2007.

Period	Organisation Model
1891 - 1937	The consortia were contracts celebrated among municipalities whose effectiveness depended upon state approval.
1937	The Constitution recognises that municipalities association in consortia are legal public entities.
1961	It is created the first Brazilian inter-federative autarchy.
1964 - 1988	Administrative consortia arise as collaboration agreements without legal personality.
From 1998	Creation of several public consortia. The Constitutional Amendment nº 19 changed the art. 24 of the Constitution of 1988, introducing the concepts of public consortia and the associated management of public services.
2005	Public Consortia Act
2007	The Decree 6,017, of 17-01-2007, regulates the Public Consortia Act

Source: Adapted from Rieiro, 2007.

As it can be seen, between 1964 and 1988 administrative consortia arose as simple collaboration agreements, without legal personality, reflecting the period of centralism of the military government. From the 1990s, based on the Constitution of 1988, a great number of public consortia appeared in Brazil, especially in the health field. Consortia were also formed around specific themes, being the most common the regional development and the environment, water resources and sanitation.

Most of the consortia established in the country involve small and medium communities. Only 5% of the consortia include cities with more than 500,000 inhabitants (Spink, 2000, apud Gouvea, 2005). Gouvea (2005) states that the main obstacle to the formation of intermunicipal cooperation is still the autarchic aspect of the Brazilian municipalities, in a 'compartmentalised' federalism context, which rigidly separates the counties. Thus, the Brazilian federative framework does not ease the cooperation between municipalities.

The discussion about the new Public Consortia Act began in August 2003, aiming to regulate the Article 241 of the Constitution and give more legal and administrative security to the partnerships among the consortium parts. In 2005, the Congress approved the new Act. The public consortia, according to this Act, are partnerships formed by two or more entities of the federation to achieve common interest goals, in any area. The consortia can discuss how to promote the regional development, manage solid and wastewaters disposal, and build new hospitals or schools. They have their origin in the municipalities associations, which were already established in the Constitution of 1937. One of the purposes of the public consortia is to enable and make viable the public administration of the metropolitan areas, where urban problems solutions require joint policies and actions.

The consortium also allows small municipalities to act in partnership and, with a gain in scale, improve their technical, operational and financial capabilities. It is also possible to make alliances in areas of common interest, such as watersheds or regional development poles, improving public services. Indeed, the new Act brought to the public scene a promising tool for the management of common problems in urban areas, offering to the public entities a viable alternative for cooperation at the supra-municipal level.

Sanitation Sector Regulation

After a long period without a regulatory mark for the sanitation services, it was approved, on 5 January 2007, the Basic Sanitation Act, n° 11,445. With this Act, the country established a modern regulatory mark for the sanitation sector, integrated with the National Policy of Water Resources Management, and establishes the national guidelines for the basic sanitation sector. This Act considers as basic sanitation the public supply of potable water services; the collection, transportation, treatment and adequate sewage final disposal; the collection, transportation, transfer, treatment and final disposal of household waste and garbage originated in the public streets and open areas; the drainage and urban storm water management, considering the conduction, detention or retention of the flood flows and the treatment and final disposal of the rainwater drained from the urban areas.

The Act states in its fundamental principles, among other things, the need of making the sanitation services available for all urban areas, preserving citizens health and public and private structures. The sanitation services need to be articulated with urban and regional policies of development and integrated with water resources management. These principles clearly demonstrate the integrative perspective of the Act, fleeing from the traditional view of the sanitation sector in the country, especially when compared with the actions of the former National Department of Sanitation Works (DNOS).

The Basic Sanitation Act also provides several innovations, among others, the possibility of the holders of public sanitation services delegate these services to Public Consortia. Another unprecedented advance is the possibility of including in the contracts some progressive and gradual goals for services expansion, increasing in quality, efficiency and rational use of water, and energy rationalisation.

Members of the Federation are also allowed to establish funds, separately or together in public consortia, which may be composed, among other resources, by a portion of the revenue from the services, in order to cover the costs of the universalisation of the public sanitation. These funds, in addition to traditional funding sources, can solve the chronic lack of financing for the sector, especially in relation to urban drainage, where resource allocation is more uncertain.

The Act consolidates the possibility of formation of the Public Consortium for providing regionalised public sanitation services, as stated in the Public Consortium Act.

The Decree n° 6,017 of 17 January 2007, that regulated the Public Consortia Act, details the way public bodies may constitute consortia. The first aspect to be noted is that the public consortium will be constituted as a legal entity formed solely by members of the Federation, organised as a public association, with legal personality under public law and autarchic nature, or as a legal entity of private law with non-profit aims.

The goal of the consortium will be determined by the associated public entities, assuming, among others, the following possibilities:

- Associated management of public services;

- Services provision, including technical assistance, structures construction and goods supply to the direct or indirect administration of the consortium members;

- Sharing or common use of the equipments or instruments, including those for management, maintenance, data processing;

- Technical studies production;

- Water resources rational use promotion and the environment protection;

- Performance of functions in the water resources management system that have been to them delegated or authorised;

- Local and regional policies and actions for urban and socio-economic development.

The prevision for the exercise of the multi-sectoral functions opens the way for the establishment of a technical agency with legal competence for integrating the public policies involving environment, water resources, sanitation and land use planning within a regional scope.

Integrated Perspectives: Water Resources, Sanitation and Urban Development

Probably, the most urgent and complex task of the agenda of public managers really committed to build a sustainable future for the cities refers to promote the integration of public policies concerning water resources, sanitation and urban land use ordering.

The metropolitan question is an issue of increasing importance, and the built environment worsens with the growing of the cities and the conurbation process. At the moment, in Brazil, there are available tools for building institutional arrangements that resume the management in metropolitan basis, replacing the model that prevailed in the last twenty years, which focused in local and fragmented policies. Thus, there are reasons to believe that the new institutional arrangements figuring in the country offer alternatives for the shared management between states and municipalities, especially in the larger urban agglomerations. The public consortia may have more political and legal legitimacy to plan the interventions that could cause better impacts in the territory, in an integrated manner, interacting with all the levels of government and society. It is strongly needed the long-term planning resumption, based on effective cooperation mechanisms.

Specifically regarding the action of the municipalities, there is a vast field of possibilities to be sought, especially after the approval of the City Statute. The new Master Plans, previewed in this statute, can and should incorporate mechanisms for a more effective management of land use, using a wider range of legal, economic and tax-oriented instruments for the urban development on a sustainable basis.

Finally, the improvement of the technical management of cities and metropolitan areas challenge remains. Again, it is stressed the need for creation of cooperative structures, not only among the various municipalities of the same metropolitan area, but also between these municipalities and the state, for the definition and implementation of policies in an integrated manner.

CASE STUDY: THE SEEK FOR SUSTAINABLE SOLUTIONS FOR URBAN FLOODS IN IGUAÇUSARAPUÍ RIVER BA-SIN AT RIO DE JANEIRO STATE, BRAZIL

A case study regarding the Iguaçu-Sarapuí River Basin, located in the western portion of the Guanabara Bay Basin, which lies in the Metropolitan Region of Rio de Janeiro is discussed. Figure 6 shows a map of this area, with the Cities that are in this basin. This is one of the most critical areas in the state regarding urban flooding. This region is densely occupied and presents great urban and industrial development areas, as well as wide rural zones in an urbanising process, and reminiscent areas of natural vegetation on the upstream reaches of the basin. This case study intends to illustrate and complement the conceptual discussion held in this chapter and to show how complex the interaction of urban drainage problems and the city growth can be in a context of an unplanned and non-integrated reality. In this region, urban expansion dynamics is, in general, marked by irregular occupation, in terms of land tenure and urban regularisation, and lack of sanitation.

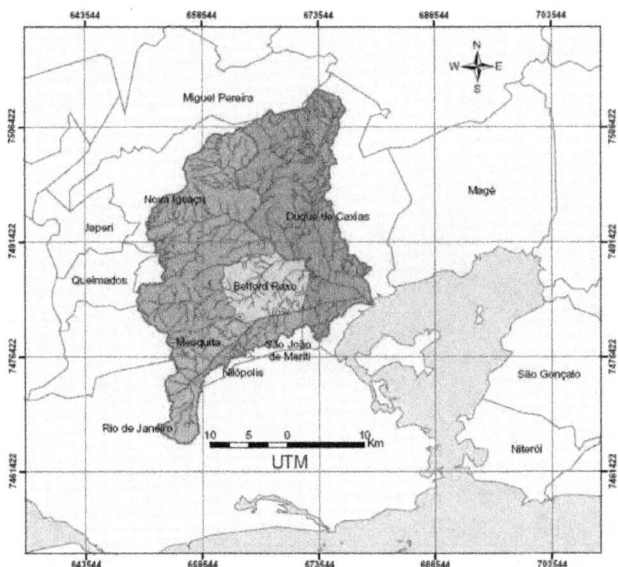

Figure 6: Iguaçu-Sarapuí River Basin at Baixada Fluminense Lowlands in the Metropolitan Area of Rio de Janeiro.

The region under consideration has a great portion of Baixada Fluminense Lowlands. Interventions for flood mitigation were supported by Federal Government on 1930's (canals construction, dams, floodgates and pumping stations). At that time, the hydraulic structures were projected for agricultural uses. A migratory process for this area began on the 1950's and accelerated from the 1970's on. In the beginning of the 1990 decade, Baixada Fluminense Lowlands sheltered more than 2 million inhabitants in 6 counties. More than 350 thousand of these inhabitants suffered the effects of significant floods. The chaotic process of urbanisation resulted in the occupation of the main rivers bed, what has made almost impossible the maintenance of the watercourses; the acceleration of the process of rivers and canals sedimentation due to deforestation of the slopes and inadequate solid waste disposal; and the increase of the runoff, due to uncontrolled vegetal removal and consequent substitution by impervious surfaces.

The basin of Iguaçu-Sarapuí River, in the past years, became a stage for articulated actions focusing on flood control and environmental recovery, in the context of the revision of its Water Resources Master Plan, in a study made by the Federal University of Rio de Janeiro for the State Institute of the Environment (INEA, in Portuguese). This study was supported by a mathematical model, called MODCEL, capable to represent the system in integrated terms (Mascarenhas & Miguez, 2002). Figure 7 shows the mapped flood conditions for present situation, represented over the modelled area of Iguaçu River Basin.

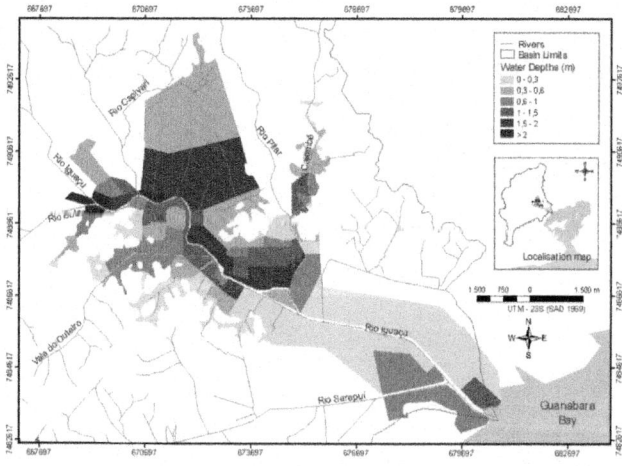

Figure 7: Flood map of Iguaçu River Basin, for a design rainfall of 20 years of return period, calculated with a mathematical model aid.

Some of the proposed actions were:

- Maintenance of spaces free of urbanisation, preventing the aggravation of flooding at the consolidated urban areas;

- Land use regulation and control, by means of the establishment of environmental preservation areas;

- Implementation of urban parks;

- Creation of public consortiums for integrated planning of policies for multi-counties interests (recognising the importance of the metropolitan planning);

- Revision and adaptation of the municipalities urban planning instruments.

Three kinds of parks were designed:

1. Fluvial Urban Park – longitudinal parks along rivers, with the purpose of protecting water course banks, also avoiding their irregular occupation by low income population.

2. Flooding Urban Park – longitudinal parks implemented in low elevation areas to allow frequent inundations, which will help to damp flood peaks.

3. Environmental Urban Park - parks with greater dimensions, flat or not, with the purpose of environmental preservation and land use valuing, aiming to minimise runoff generation and maintaining a buffer of pervious surfaces.

Figure 8 shows the general propositions for Iguaçu River Basin, while Figure 9 shows two more detailed examples of the proposed parks.

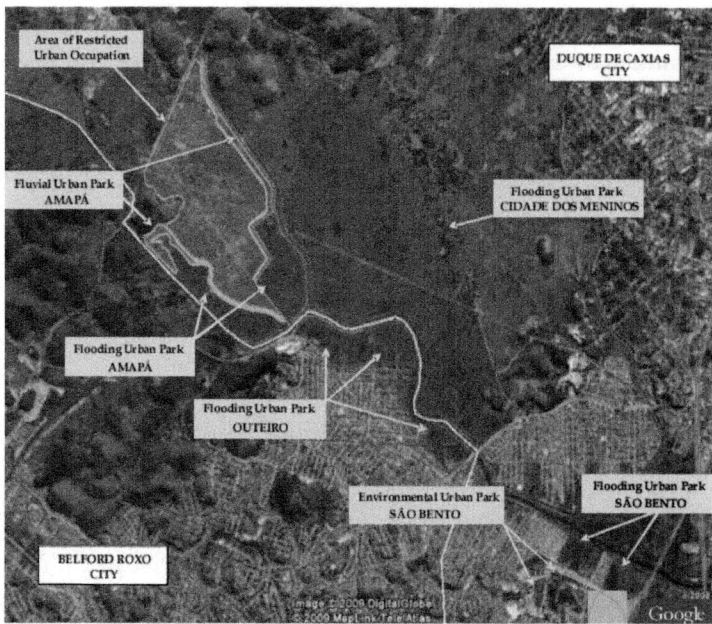

Figure 8: Flood control measures proposed for Iguaçu River Basin.

Complementary actions held by the State include the articulation with every Municipality in the basin, in order to implement the proposed measures, create local conditions for urban land uses control and develop environmental education campaigns, with the financing of the Federal Government, through a specific Program of Developing Acceleration.

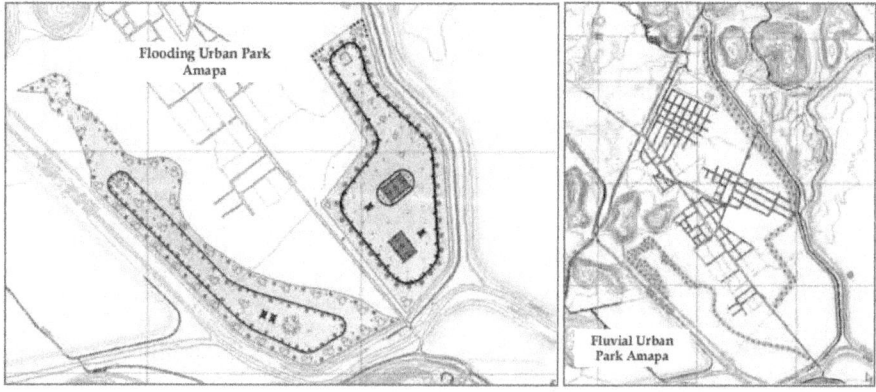

Figure 9: Fluvial Urban Park and Flooding Urban Park examples.

According to INEA information, these interventions will benefit directly and indirectly, around 3 million inhabitants of Baixada Fluminense and, encompassing: the recovery of 80 km of degraded riverbanks, promoting the resettlement of 2,200 families from risk areas to new housing developments at neighbouring areas; the implementation of parks and recreational areas to protect the recovered riverbanks from new occupations and for storing temporarily floods; the definition of new areas for environmental preservation purposes; the construction or recovery of approximately 70 km of streets along the areas of intervention, as well as the recovery of narrow crossings, bridges, aqueducts, gates and polders. It is also estimated the planting of more than 200,000 trees along the riverbanks. In order to sensibilise and mobilise the local communities for social and environmental problems and the importance of participation and social control, INEA has been developing the process of monitoring and evaluating this project through local committees and through the regional forum of participation and social control.

CONCLUSION

City growth is a world trend and sustainability is a central point to be considered in the next times to come. In a general way, however, great cities present lots of problems to deal with: land use control needs, sub-habitation, unemployment, poverty, inefficient transportation, insufficient public services, lack of infrastructure, among others. The question of water resources management and sanitation aspects are of fundamental importance in this scenery. The urban flood problem is certainly one of the most important challenges that cities will have to face. The urbanisation process is one of the man-made actions that most affect floods. On the other side, in the context of a city, the flood process is one of the facts that most degrades it. Considering urban drainage in the context of the integrated city development, however, the sustainability perspective opens a diversified set of opportunities to be explored as integrated solutions, in the fields of hydraulic engineering, architecture and urbanism, city planning and management, social disciplines and economy concerns.

Acknowledgments

The authors acknowledge CNPq for the scholarships and for financing their researches. Also, they would like to highlight that this research is part of the project SERELAREFA (Semillas REd LAtina Recuperación Ecosistemas Fluviales y Acuáticos) in the context of the Program FP7 IRSES PEOPLE 2009.

REFERENCES

1. Andoh, R. Y. G. & Iwugo, K. O. (2002). Sustainable Urban Drainage Systems: - A UK Perspective. Proceedings of the 9th International Conference on Urban Drainage, Portland, Oregon, USA, 2002.

2. Arzet K. (2010). O Rio Isar; Munique, Alemanha, In: Revitalização de Rios no Mundo: América, Europa e Ásia. Machado A.T.G.M., Lisboa A.H., Alves C.B.M., Lopes D.A., Goulart E.M.A., Lite F.A., Polignano M.V., pp 153-168, Instituto Guaicuy, ISBN 978.85.98659.08.4, Belo Horizonte, Brazil. (in Portuguese)

3. Batista, M.; Nascimento, N. & Barraud, S. (2005). Técnicas Compensatórias em Drenagem Urbana, ABRH, ISBN 858868615-5, Porto Alegre, Brazil.

4. Benevolo, L. (2001). História da Cidade, Editora Perspectiva, São Paulo, Brazil.

5. Biswas, A.K. (1970). History of Hydrology, North Holland Publishing Company, Amsterdam.

6. Brazil. (1979). Federal Act 6,766, of December 19, 1979. Federal Official Gazette of Brazil, Brasília, DF, 20 December. 1979. Section 1, Brasília, Brazil. (in Portuguese)

7. Brazil. (1997) Federal Act 9,433, of January 8, 1997. Federal Official Gazette of Brazil, Brasília, DF, n. 6, 09 January. 1997. Section 1, Brasília, Brazil. (in Portuguese)

8. Brazil. (2001) Federal Act 10,257, of July 10, 2001. Federal Official Gazette of Brazil, Brasília, DF, n. 133, 11 July. 2001. Section 1, Brasília, Brazil. (in Portuguese)

9. Brazil. (2007) Federal Act 11,445, of January 5, 2007. Federal Official Gazette of Brazil, Brasília, DF, n. 8, 11 January. 2007. Section 1, Brasília, Brazil. (in Portuguese)

10. Burian, S.J. & Edwards, F.G. (2002). Historical perspectives of urban drainage, Global Solutions for Urban Drainage, Proceedings of the 9th International Conference on Urban Drainage, Portland, September 2002.

11. Canholi, A.P. (2005). Drenagem urbana e controle de enchentes, Oficina de Textos, ISBN 8586238430, São Paulo, Brazil.

12. Carneiro, P. R. F. (2003). Dos Pântanos à Escassez: uso da água e conflito na Baixada dos Goytacazes, Annablume, São Paulo, Brazil.

13. CEPAL – Comisión Económica para América Latina y el Caribe. (1999). Tendencias actuales de la gestión del agua en América Latina y el Caribe (avances en la implementación de las recomendaciones contenidas en el

capítulo 18 del Programa 21. LC/L.1180, Santiago de Chile.

14. Chocat, B., Krebs, P., Marsalek, J., Rauch, W. & Schilling W. (2001). Urban Drainage Redefined: from Stormwater Removal to Integrated Management. Water Science and Technology, Vol. 43, No. 5, (2001), pp. (61–68).

15. CIRIA (2007). The SUDS Manual, by Woods-Ballard, B.; Kellagher, R.; Martin, P.; Bray, R.; Shaffer, P. CIRIA C697.

16. Coffman, L.S., Cheng, M., Weinstein, N. & Clar, M. (1998). Low-Impact Development Hydrologic Analysis and Design. Proceedings of the 25th Annual Conference on Water Resources Planning and Management, Nova York, USA, 1998.

17. Dourojeanni, A. & Jouravlev, A. (1999). Gestión de cuencas y ríos vinculados con centros urbanos. CEPAL - Comisión Económica para América Latina y el Caribe, 1999.

18. Dourojeanni, A. & Jouravlev, A. (2001). Crisis de Gubernabilidad en la Gestión del Agua. Serie Recursos Naturales e Infraestructura, No. 35, Cepal, División de Recursos Naturales e Infraestructura, Santiago.

19. González del Tánago, M. & García de Jalón, D. (2007). Restauración de ríos. Guía metodológica para la elaboración de proyectos. Ministerio de Medio Ambiente, Madrid, España. (in Spanish)

20. Gouvêa, R. G. (2005). A questão metropolitana no Brasil, Editora FGV, Rio de Janeiro.

21. Gusmaroli, G.; Bizzi, S. & Lafratta, R. (2011) L'approccio della Riqualificazione Fluviale in Ambito Urbano: Esperienze e Opportunittà. Anais do 4° Convegno Nazionale di Idraulica Urbana, Veneza, Itália, June 2011.

22. Hill, R. (2010) O Rio Tâmisa: Londres, Inglaterra. Machado A.T.G.M., Lisboa A.H., Alves C.B.M., Lopes D.A., Goulart E.M.A., Lite F.A. and Polignano M.V. (Org.) In: Revitalização de Rios no Mundo: América, Europa e Ásia. Machado A.T.G.M., Lisboa A.H., Alves C.B.M., Lopes D.A., Goulart E.M.A., Lite F.A., Polignano M.V., pp 131- 152, Instituto Guaicuy, ISBN 978.85.98659.08.4, Belo Horizonte, Brazil. (in Portuguese)

23. Holz, J. & Tassi, R. (2007). Usando Estruturas de Drenagem Não Convencionais em Grande Áreas: O Caso do Loteamento Monte Bello. Proceedings of the XVII Simpósio Brasileiro de Recursos Hídricos, São Paulo, Brazil, November 2007.

24. Johnson, W.K. (1978). Physical and economic feasibility of nonstructural

flood plain management measures, Institute for Water Resources, U. S. Army Corps of Engineers, Fort Belvoir, VA.

25. Jormola, J. (2008). Urban Rivers, In: Proceedings of the 4th ECRR Conference on River Restoration, Venice S. Servolo Island, Italy, June 2008.

26. Jouravlev, A. (2003). Los municipios y la gestión de los recursos hídricos. Serie Recursos Naturales e Infraestructura. CEPAL - Comisión Económica para América Latina y el Caribe, n° 66, 2003.

27. Langenbach, H.; Eckart, J. & Schröder, G. (2008). Water Sensitive Urban Design – Results and Principles. Proceedings of the 3rd SWITCH Scientific Meeting, Belo Horizonte, Brazil, 2008.

28. Lowbeer, J. D., Cornejo, I. K. (2002). Instrumentos de gestão integrada da água em áreas urbanas. Subsídio ao Programa Nacional de Despoluição de Bacias Hidrográficas e estudo exploratório de um programa de apoio à gestão integrada. USP/Núcleo de Pesquisa em Informações Urbanas - Convênio FINEP CT-HIDRO.

29. Mascarenhas, F.C.B. & Miguez, M.G. (2002). Urban Flood Control through a Mathematical Cell Model. Water International, Vol. 27, N° 2, (June 2002), pp. (208-218).

30. Miguez, M.G., Mascarenhas, F.C.B., Magalhães, L.P.C. (2007). Multifunctional Landscapes for Urban Flood Control In Developing Countries. International Journal of Sustainable Development and Planning, Vol. 2, N°2 (2007), pp. 153-166, ISSN 1743- 7601.

31. Miguez, M.G., Magalhães, L.P.C. (2010). Urban Flood Control, Simulation and Management: an Integrated Approach. In: Methods and Techniques in Urban Engineering, Armando Carlos de Pina Filho & Aloísio Carlos de Pina, pp. (131-160), In-Tech, ISBN 978-953- 307-096-4, Vukovar, Croatia.

32. Ministério das Cidades. (2004). Manual de Drenagem Urbana Sustentável, Brasília, Brazil, 2004.

33. Noh, S.H. (2010). Rio Cheonggyecheon: Seul, Coreia do Sul. Machado A.T.G.M., Lisboa A.H., Alves C.B.M., Lopes D.A., Goulart E.M.A., Lite F.A. and Polignano M.V. (Org.) In: Revitalização de Rios no Mundo: América, Europa e Ásia. Machado A.T.G.M., Lisboa A.H., Alves C.B.M., Lopes D.A., Goulart E.M.A., Lite F.A., Polignano M.V., pp 291-304, Instituto Guaicuy, ISBN 978.85.98659.08.4, Belo Horizonte, Brazil. (in Portuguese)

34. Palmer, M.A., Bernhardt, E.S., Allan, J.D., Lake, P.S., Alexander, G., Brooks, S., Carr J., Clayton, S., Dahm, C.N., Follstad Shah, J., Galat, D.L.,

Loss, S.G., Goodwin, P., Hart, D.D., Hassett, B., Jenkinson, R., Kondolf, G.M., Lave, R., Meyer, J.L., O'Donnel, T.K., Pagano, L. & Sudduth, E. (2005). Standards for ecologically successful river restoration. Journal of Applied Ecology, Vol. 42, N° 2, (April 2005), pp. 208-217.

35. Peixoto, M. C. D. (2006). Expansão urbana e proteção ambiental: um estudo a partir do caso de Nova Lima/MG. In: Novas Periferias Metropolitanas – A expansão metropolitana em Belo Horizonte: dinâmica e especificidades no Eixo Sul, Heloisa Soares de Moura Costa (organizadora); Geraldo Magela Costa, Jupira Gomes de Mendonça, Roberto Luis de Monte-Mór (colaboradores); [Editor: Fernando Pedro da Silva] – Belo Horizonte: C/Arte, 2006.

36. Pompêo, C. A. (1999). Development of a state policy for sustainable urban drainage. Urban Water, Vol. 1, N° 2, (July 1999), pp. 155-160.

37. Pompêo, C.A. (2000). Drenagem Urbana Sustentável. Revista Brasileira de Recursos Hídricos, Vol. 5, N° 1, (January-March 2000), pp. 15-23.

38. Rezende, O. M. (2010). Avaliação de medidas de controle de inundações em um plano de manejo sustentável de águas pluviais aplicado à Baixada Fluminense. MSc. Thesis, Federal University of Rio de Janeiro, Brazil, 2010.

39. Righetto, A.M., Moreira. L.F.F. & Sales, T.E.A. (2009). Manejo de Águas Pluviais Urbanas. In: Manejo de Águas Pluviais Urbanas, Projeto PROSAB, Righeto, A.M., pp. (19-73), ABES, Natal, Brazil.

40. Rieiro, W. A. (2007). Cooperação Federativa e a Lei de Consórcios Públicos. Brasília – DF: CNM.

41. Silva, R. T. & Porto, M. F. A. (2003). Gestão urbana e gestão das águas: caminhos da integração. Estudos Avançados, Vol. 47, No. 17, (January-April 2003), pp.(129-145).

42. Stahre, P. (2005). 15 Years Experiences of Sustainable Urban Storm Drainage in the City of Malmo, Sweden, Proceedings of World Water and Environmental Resources Congress, Alaska, May 2005.

43. Tucci, C.E.M. (1995). Inundações Urbanas, In: Drenagem Urbana, Tucci, C.; Porto, R.; Barros, M., pp. 15-36, Editora da Universidade/ABRH, ISBN 8570253648, Porto Alegre, Brazil.

44. Tucci, C. E. M. (2004). Gerenciamento integrado das inundações urbanas no Brasil. Rega/Global Water Partnership South América, Vol. 1, No. 1, (January-June, 2004), pp. (59-73).

45. United States Environmental Protection Agency – US EPA. (2004). the Use of Best Management Practices (BMPs) in Urban Watersheds, by

Muthukrishnan, S.; Madge, B.; Selvakumar, A.; Field, R.; Sullivan, D. EPA/600/R-04/184S.

46. Wong, T.H.F. (2006). Water Sensitive Urban Design – the Journey Thus Far. Australian Journal of Water Resources, Vol. 10, No. 3, (2006), pp. 213-222.

Chapter 3

CASE STUDY: HYDRAULIC MODEL EXPERIMENT TO ANALYZE THE HYDRAULIC FEATURES FOR INSTALLING FLOATING ISLANDS

Sanghwa Jung, Joongu Kang, Il Hong, Hongkoo Yeo

Department of Water Resources Research & Environment Research, Korea Institute of Construction Technology, Ilsan, Korea

ABSTRACT

The viewpoint of a river is changing as people regard the river as water-friendly space where they can enjoy and share the space beyond the simple purpose of flood control alongside the improving social level. The floating islands installation was planned featuring three islands. The river's flow and channel stability could be changed when new structures are built in a river. Hence an analysis of the hydraulic characteristic changes should need. The hydraulic model experiment in this study sought to review the impacts of the floating islands installation on the safety of flood control and stability of river channel. This study analyzed the hydraulic features affecting the surrounding stability when installing floating islands and proposed stable floating islands layout in terms of hydraulics based on the experiment results.

INTRODUCTION

The viewpoint of a river is changing as people regard the river as water-friendly space where they can enjoy and share the space beyond the simple purpose of flood control alongside the improving social level. Floating islands installation was carried out according to the plan to shape the cultural space along the river, i.e., to provide water culture and leisure space. Floating islands were planned featuring three islands. The hydraulic model experiment (HME) in this study aimed at reviewing the impact of floating islands installation on the safety of flood control and stability of river channel. The main review items were flow duration, velocity, and shear stress.

The experiment target area is the downstream area of Han River located at the center of Seoul. The project section is an approximately 3.5 km-long straight line downstream between Hannam Bridge and Dongjak Bridge. The section marked with the red dotted line in Figure 1 is the project target section, and the blue dotted line-marked section, the section of main interest with some changes in the target section.

The target area's flood frequency is high due to the high altitude of the water front of Han River and owing to damage by floods occurring at least 3 times annually. This area was used as grassland and green land-centered area because of the frequent floods. The basic plan of the target area within Han River consists of educational facilities construction including ecosystem natural learning arena and complex cultural space shaping as an event plaza and view plaza through riverside improvement. The basic plan is established to shape the ecosystem river bank and for nature-friendly or water-friendly river bank protection in terms of the change of bank protection. Moreover, in the Banpo zone, a plan to create cultural space has been established through the floating islands installation. Consisting of 3 islands in Figure 2, the planned floating islands shall be installed on the left of the straight downstream section of Banpo Bridge as water cultural and leisure space in a 9200 m^2 area. Here, two issues are flood control stability and hydraulic movement. First is topographical change on the left riverside of Han River in the target area including change in low?

Figure 1: Project target section (red dotted line) and section of main interest (blue dotted line).

Figure 2: View before and after the installation of floating islands.

Water channel cross section according to change of bank protection shape. The impacts of such topographical change on flood water level and current field should be reviewed. Second is floating islands installation. A review of the impacts of the floating islands installation on flood water level and riverbed change should be conducted, including a review of the surrounding flow duration when installing the floating islands to check whether a disadvantageous hydraulic phenomenon occurs. This paper presents a design proposal that can minimize hydraulic change within the river channel based on floating islands installation.

THEORY AND SCALE

The hydraulic model experiment (HME) was carried out to find hydraulic solutions to phenomena limited to specific areas including the construction of hydraulic structure and river improvement. As such, HME has various purposes according to project features, and it should consequently be carried out with clear purposes. Likewise, a variety of plans including experiment size in line with the purposes need to be established. The model and actual-sized prototype in the HME should meet three similitudes: geometric similitude, kinematic similitude, and dynamic similitude [1]. In the geometric similitude, the length ratio needs to be constant at the corresponding point of the actual-sized prototype and model. In the kinematic similitude, the ratio of velocity and acceleration needs to be constant at the corresponding point of the actual-sized prototype and model. In the dynamic similitude, the ratio of strength needs to be constant at the corresponding point of the actual-sized prototype and model. The element that becomes a criterion in the law of geometric similitude is expressed as $L_r = L_p/L_m$ (L: length, with r, p, and m representing the scale ratio, actual-sized prototype, and model, respectively). The velocity and acceleration ratio is expressed as $a_r = U_r/T_r = L_r/T_r^2$. Here, time ratio is expressed as T_r, $T_r = T_p/T_m$.

The dynamic similitude is the law matching the ratio of external forces that add to or reduce the flow of fluid and inertia force of fluid. Here, the main external forces are gravity, viscosity, surface tension, and pressure. When carrying out HME, similitude is decided by a main external force. In this study, the main external force was gravity because of the open river channel condition; thus, the Froude similitude as the ratio of gravity and inertia force was used. The main physical elements are presented as follows:

$$\frac{\text{Inertiaforces}}{\text{Gravity}} = \frac{\rho L^2 U^2}{\rho g L^3} = \frac{U^2}{gL} = Fr^2$$

$$U^2 = \frac{L_r}{T_r} = \sqrt{g_r L_r}, \quad U_r = \sqrt{L_r}, \quad T_r = \sqrt{L_r}$$

The scale of the hydraulic model is decided by comprehensively reviewing the reproduction ability of the actual-sized prototype, lab's discharge supply capability, experiment model production space, and ease of measurement. Generally, the length is much greater compared with the depth in a river model. Therefore, a large river tends to have wide river channel compared with depth despite the smaller length. As such, distorted scale is inevitable when carrying out large river HME. In this experiment, the change of flood water level according to the conditions was one of the most important items. For this reason, the scale was determined in the maximum size allowed by the impact experiment supply discharge and experiment site conditions while securing maximum depth and minimizing distortion considering the accuracy of water level measurement. The model used in this experiment has 1/120 horizontal scale, 1/50 vertical scale, and 2.4 distortion. The conversion ratios of the hydraulic volume of HME are presented in Table 1.

Table 1: Scales and conversion ratios of fixed bed hydraulic model experiment.

Conversion ratio of hydraulic volume	Conversion formula	Scale
Horizontal length scale. X_r	X_r	120
Vertical length scale. Y_r	Y_r	50
Area ratio. A_r	$X_r Y_r$	6000
Velocity ratio. V_r	$Y_r^{1/2}$	7.07
Discharge ratio. Q_r	$X_r Y_r^{3/2}$	42426.41
Slope ratio. S_r	X_r / Y_r	0.42
Coefficient ratio of intensity of illumination. n_r	$X_r^{-1/2} Y_r^{2/3}$	1.24

MODEL PRODUCTION AND MEASUREMENT METHOD

Model Production

For the model production, the model standard points were set by performing model measurement using a total station after completing the arrangement work on the model installation space. Along the set model standard points, the outer wall was set up using blocks, and the exterior was finished with mortar. In this manner, topographical installation work was executed. For the model section, the topographical plate model was manufactured by measurement line by cutting plywood according to the riverbed section type including bank, low water channel, and terrace land of the river after printing out as real photo through the conversion of riverbed cross section acquired from actual river measurement into a model scale. Concerning the produced topographical plates, filling work was done by paving with sand after placing by measurement line through model measurement. The fixed bed model was produced by forming paved filling materials using cement mortar along the reproduced riverbed section on the plywood after flushing with water. The produced topographical plates were installed in the precise locations by measuring with the total station. By referring to the ground plan and topographical map in Figure 3, the sections were precisely connected to reproduce the same topographical shape as the real one as much as possible. To minimize the discharge water level of the model, the water level control floodgate downstream was produced and installed as a blind type in the horizontal direction. The bridges in the model were produced with acrylic in line with the scale in terms of the shapes confirmed through onsite study along with bridge plans. The models of the three floating islands used for this experiment were made with styrofoam and these were installed to maintain 2 m draft in Figure 4. This experiment needed large discharge.

Figure 3: Target area model.

Figure 4: Floating islands installation.

Thus, more than the maximum discharge supply of 0.9 m³/s was guaranteed using three pumps. The supplied discharge was measured using the highly reliable weir mode and was calculated according to a discharge formula and the discharge calibration curve. A weir tank (3 m wide) was used as discharge supply tank.

Measurement Method

For the measurement of the starting point's water level and depth, a digital water level gauge (PH-355, KENEK) with 0.1 mm measurement accuracy was used. For the measurement of velocity, 1D propeller velocity gauge (VO-1000, KENEK) was used with regard to the entire HME target area's measurement line. In the installation section of the floating islands, a 3D electronic velocity gauge (VM-1001, KENEK) that can acquire data of 100 Hz per second was used. To identify the flow impacts around the floating islands, velocity was measured using LSPIV (Large-Scale Particle Image Velocimetry) for surface flow. For the measurement of flow duration, observation was done using red pigment to identify precisely the flow duration in the river; the main parts were observed through sketch, photographing, and videotaping.

LSPIV in **Figure 5** is used to measure the velocity of each particle by measuring the particles' displacement in the images after comparing two images shot with interval in time. The advantage of LSPIV is that it can be applied to large indoor experiment or site with just simple devices and software for image processing and analysis [2, 3]. LSPIV shoots the target area using a video camera and focuses on processing the images using a software program. LSPIV is a similar method of the conventional particle image velocimetry (PIV) technique [4,5]. There are two major differences between PIV and LSPIV. First, LSPIV measures generally larger fields of view than conventional PIV. Second, LSPIV needs inexpensive equipment of illumination and recording device to analyze surface flow [6]. And this technique does not have restrictions in laser examination and camera arrangement essential to the existing PIV. As a disadvantage of LSPIV, however, it is difficult to acquire 2D or 3D current field information in water because there is no laser examination procedure. In some cases, LSPIV acquires distorted images; thus, correlation analysis should be performed after the correction work of the shot images. For this reason, accuracy drops compared to the existing PIV. Despite such demerit, the reason LSPIV attracts interest is that the measurement of the entire current field is possible at much lower expenses and with less efforts and within a short time compared with the existing use of equipment and manpower in large-scale experiments or site survey.

HYDRAULIC MODEL EXPERIMENT

Experiment Conditions

The purpose of HME of floating islands is to conduct an experiment in the fixed bed model, review the safety of flood control through flood water level, velocity, and flow duration measurement in the Banpo zone, analyze the change impacts, and reflect the experiment results on the project plan in the Banpo zone. As shown in **Table 2**, four experiment conditions were demonstrated, i.e., two cases each before and after the installation of the floating islands. Flood conditions by frequency were presented in the basic river improvement plan. The discharge condition of flood on the riverside around the BanpoBridge was the value calculated through the 1D numerical value model Hec-Ras for the flood water level on the riverside around the BanpoBridge area. The measurement of water level and velocity was carried out at 14 measurement lines and 237 measurement points within the 3.5 km section (**Table 2**). The measurement lines and points were shown in Figures 6 and 7.

Analyses of Hydraulic Characteristics Features before and after the Floating Islands Installation

HME of floating islands seeks to identify the safety of flood control on the design proposal of the floating islands. Actually, 7066 m³/s was applied as riverside flood water discharge in the Banpo zone along with frequency of 200 years after reproducing the floating islands with models.

Before and after the installation of the floating islands, the average water level difference by measurement line

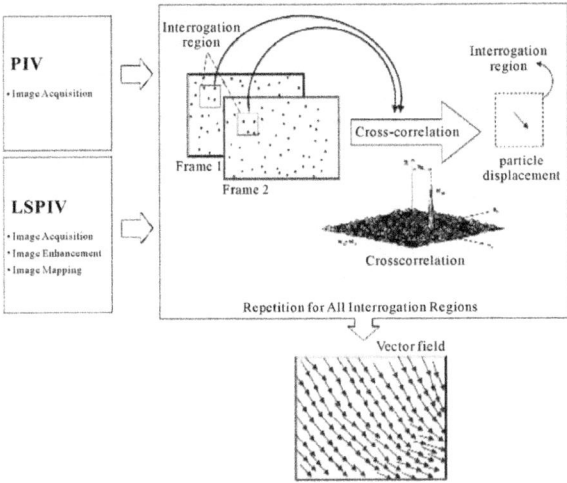

Figure 5: Analysis process using LSPIV.

Table 2: Experiment conditions.

Experiment case	Frequency	Experiment discharge (m³/s)		Beginning point's water level (EL. m)
		Actual-sized prototype	Model	
1 (Before installation)	Frequency of 200 years (planned flood water amount)	37.000	0.872	15.37
2 (After installation)				
3 (Before installation)	Riverside flood discharge in the Banpo zone (actual measurement data at Han River Bridge)	7066	0.167	6.53
4 (After installation)				

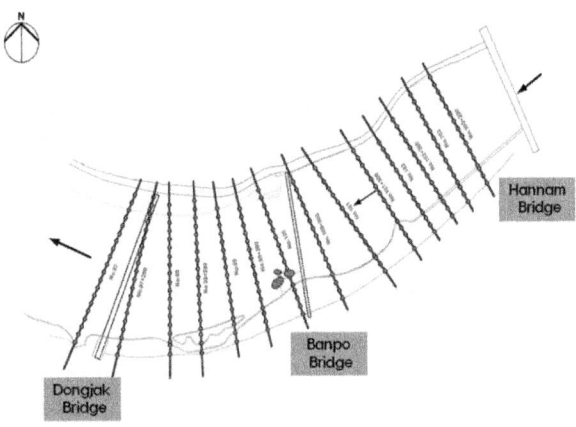

Figure 6: Measurement point location chart of hydraulic model experiment.

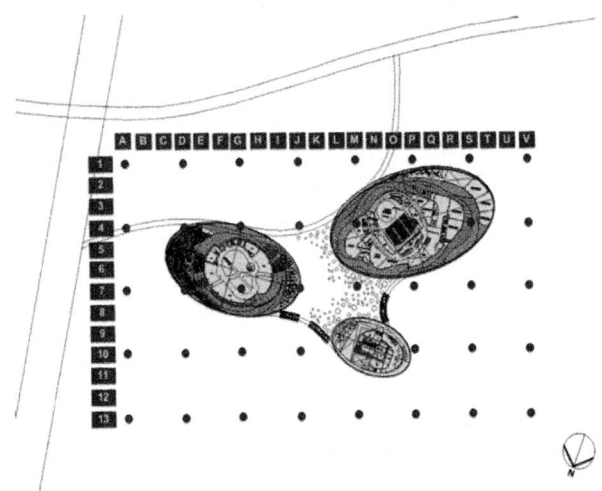

Figure 7: Velocity measurement location chart around the floating islands.

Was a maximum of 0.01 m and a minimum of –0.01 m in the case of frequency of 200 years? As a result, the change of flood water level was minimal in Figure 8. To review the actual impacts on the safety of flood control, a review of flood water level on the left and right sides of the river was conducted. The upstream section of Banpo Bridge is a relatively straight line section; thus, no clear

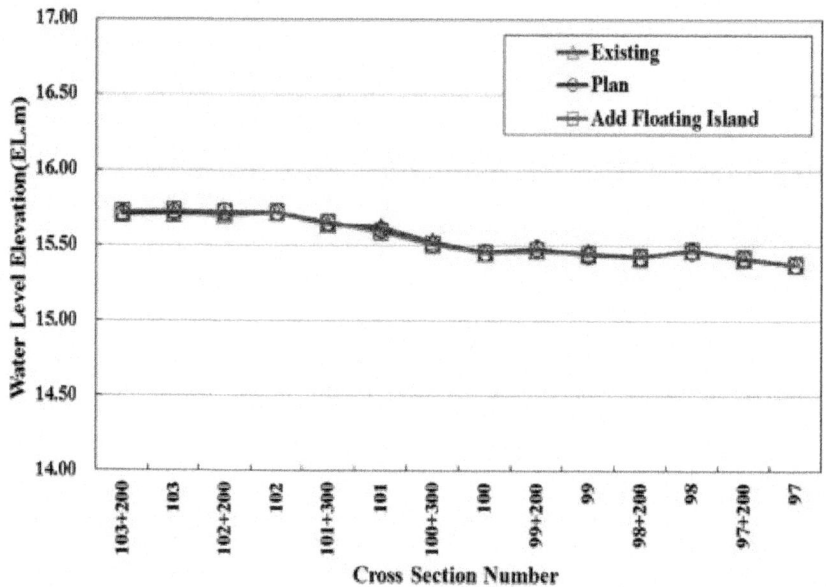

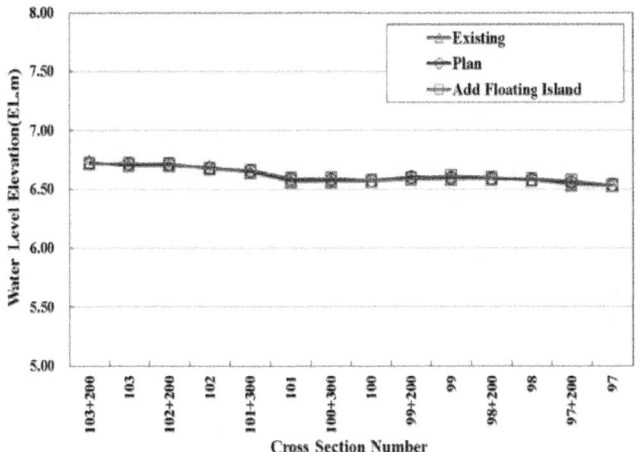

Figure 8. Comparison of water level between the current and planned cross sections and floating islands (Frequency of 200 years and bankfull discharge).

Water level difference was exhibited on the left and right riversides, but the water level on the left of the river was higher than the right because of the curve part and whirlpool formation around Seorae Island downstream around Banpo Bridge. Although there are points wherein the water level goes up locally on the left side of the river, it is just local phenomenon arising from the whirlpool, hardly affecting flood control safety; hence the minimal change scope of flood water level. In addition, there seems to be no problem in flood control stability.

The cross sectional average velocity before and after floating islands installation was measured to be a maximum of 0.06 m/s and a minimum of –0.11 m/s; average riverside flood discharge was a maximum of 0.10 m/s and a minimum of –0.01 m/s, with minimal differences in Figure 9. In the case of frequency of 200 years in the section of measurement points 5 - 9 at measurement line no. 100 where the floating islands were installed, cross sectional average velocity was a maximum of 0.02 m/s and a minimum of –0.12 m/s, and riverside flood discharge was a maximum of 0.01 m/s and a minimum of –0.01 m/s. Although the impacts on the river channel were minimal, local flow change surrounding the floating islands was observed.

Figures 10 and 11 show the flow duration through images acquired by the LSPIV experiment. Some drift was observed following the installation of floating islands; the installation was confirmed not to affect the surrounding flow, however. In case floating islands were installed, the flow duration had no change around the upstream area of BanpoBridge, but the phenomenon

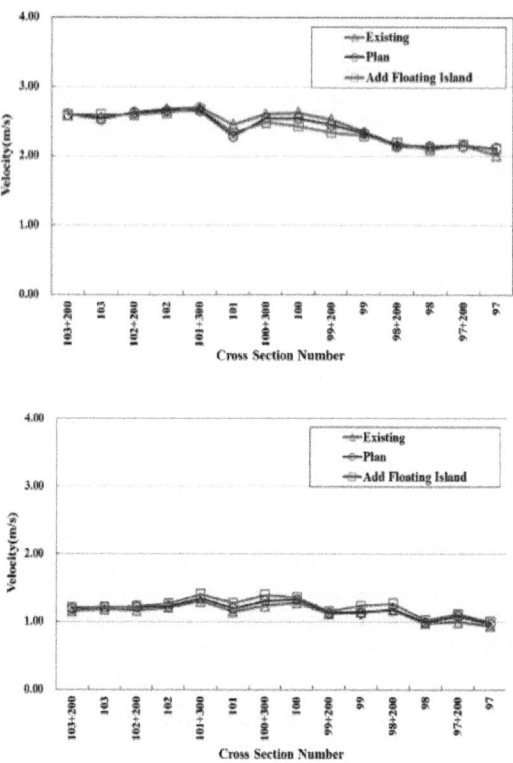

Figure 9. Comparison of velocity of floating islands and current and planned cross sections (Frequency of 200 years and bankfull discharge).

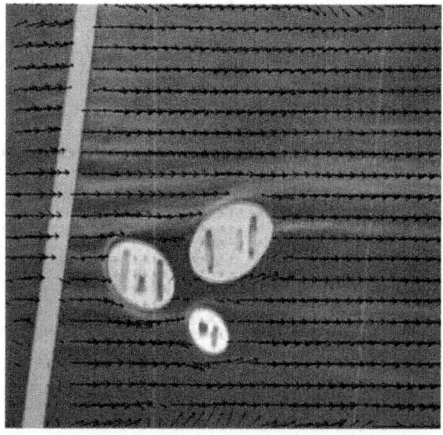

Figure 10. Surface current field after floating islands were installed with frequency of 200 years.

Figure 11. Surface current field after floating islands were installed with bankfull discharge.

Wherein flow is divided around the floating islands downstream of Banpo Bridge occurred. Nonetheless, strong whirlpool or backwash did not occur, with the impact scope downstream observed within a maximum of 150 m from the end of the installed floating islands. The area affected by the floating islands was about 400 m × 200 m including the floating islands. In the case of riverside flood discharge in the Banpo area (7066 m^3/s), small whirlpool and stagnant area were observed in some sections due to topographical impact on the left side of the installed floating islands area before the floating islands were installed. When the floating islands were installed, however, velocity increased owing to the impacts of the floating islands on the current field on the left side of the floating islands. Such velocity increase was observed to increase backwash slightly in the stagnant area and small whirlpool, which occurred because of topographical impacts prior to the floating islands installation. Concerning the surrounding water level change arising from the floating islands installation, water level rise was locally observed owing to the clash between flow and floating islands, but no water level rise impact was observed based on the water level comparison at the upstream measurement lines. The water level impact resulting from the clash between flow and floating islands is a local phenomenon; there was no big difference by flood frequency as observed within approx. 5 - 10 m section upstream of the floating islands.

Bottom Velocity Distribution and Shear Stress Surrounding the Floating Islands

The analysis of changes in bottom velocity and shear stress surrounding the floating islands was carried out to predict riverbed change that can be caused by the floating islands installation. Concerning the bottom layer's velocity change, velocity was 1.73 m/s - 2.15 m/s before the floating islands installation and 1.70 m/s - 2.18 m/s after the floating islands installation in the velocity measurement section. In other words, velocity change was minimal in the case of frequency of 200 years. Velocity change tended to diminish in the upstream area of the floating islands but tended to increase in the area where the floating islands were installed. Velocity increase in the area where floating islands were installed was about 0.1 m/s. The maximum velocity increase at the bottom layer was 0.14 m/s at D7, with maximum velocity increase of about 0.16 m/s at P1 in the measurement section. Maximum velocity decline was measured to be –0.23 m/s at A4 in **Figure 12**. When riverside flood discharge was 7066 m^3/s, velocity distribution in the velocity measurement section surrounding the floating islands was 0.50 m/s - 1.28 m/s before the floating islands installation and 0.71 m/s - 1.46 m/s after the floating islands installation. Velocity change tended to decline at measurement line G upstream section surrounding the floating islands but tended to increase at the measurement line G downstream and riverside area on the left side of the installed floating islands.

Velocity increase in the floating islands installation area was measured to be about 0.05 m/s - 0.24 m/s, and maximum velocity increase in the measurement section was about 0.026 m/s. The area where velocity increase is huge in the riverside flood discharge was mostly on the left side of the floating islands, apparently due to the reduction of discharge capacity caused by the floating islands installation in **Figure 13**.

In the case of frequency of 200 years, the distribution of shear stress before and after the floating islands installation was measured to be 1.56 kg/m^2 - 2.18 kg/m^2 and 1.61 kg/m^2 - 2.16 kg/m^2, respectively in **Figure 14**. Although maximum shear stress change existed, overall distribution difference was minimal depending on the measurement points.

Shear stress affecting the riverbed was biggest in the case of frequency of 200 years, but the increase arising from the floating islands installation does not seem to impact the riverbed since it is minimal and shear stress under floating island in the case of 7066 m^3/s was increased bigger than other discharge conditions in **Figure 15**. The point where the biggest increase was measured is the riverside bank; the increase was relatively huge because shear stress was very small as a stagnant section before the floating islands installation. Thus, the impact on the riverbed is judged to be minimal. Nonetheless, scrupulous review and measure to reduce impacts are considered necessary since flow duration in the river bank may become slightly unfavorable owing to the floating islands installation.

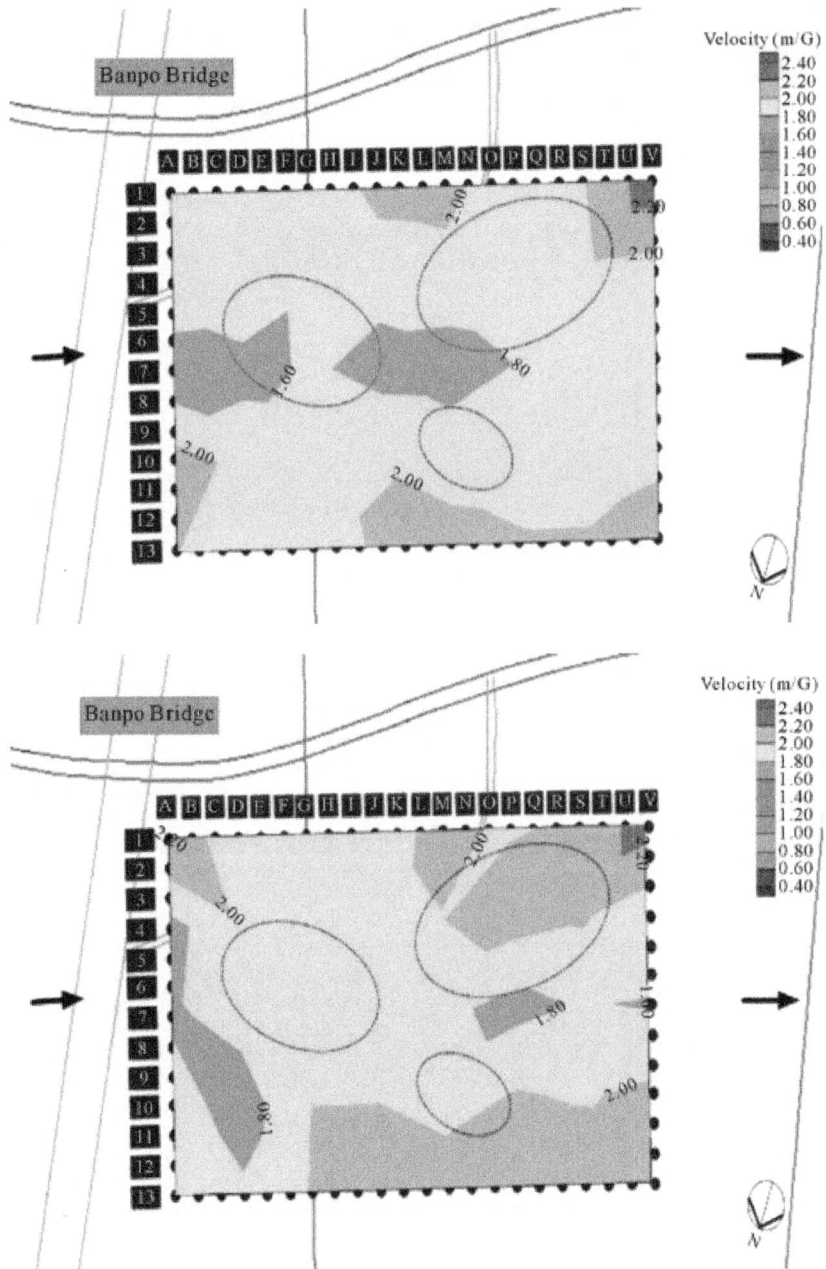

Figure 12: Velocity distribution before and after floating islands (frequency of 200 years).

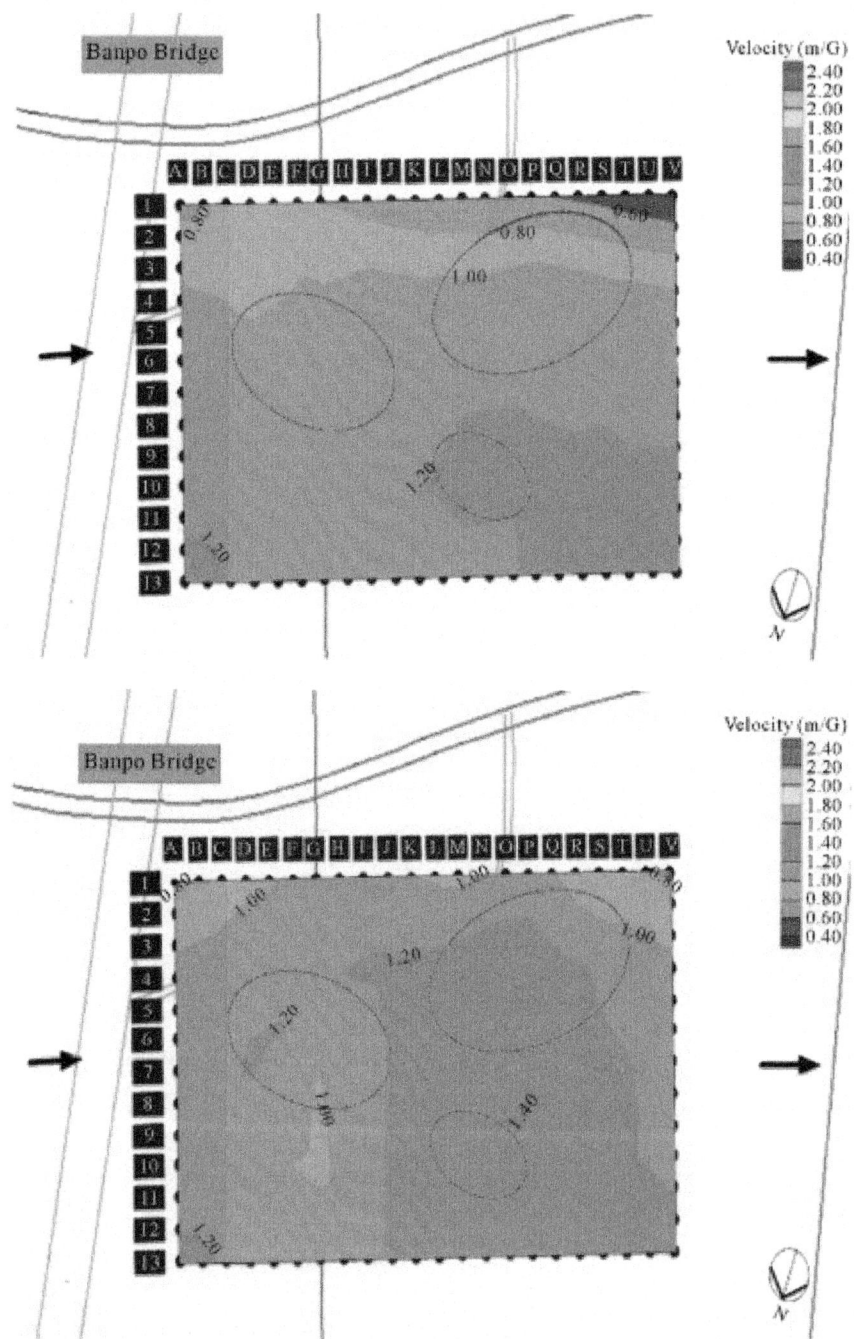

Figure 13: Velocity distribution before and after floating islands (bankfull discharge).

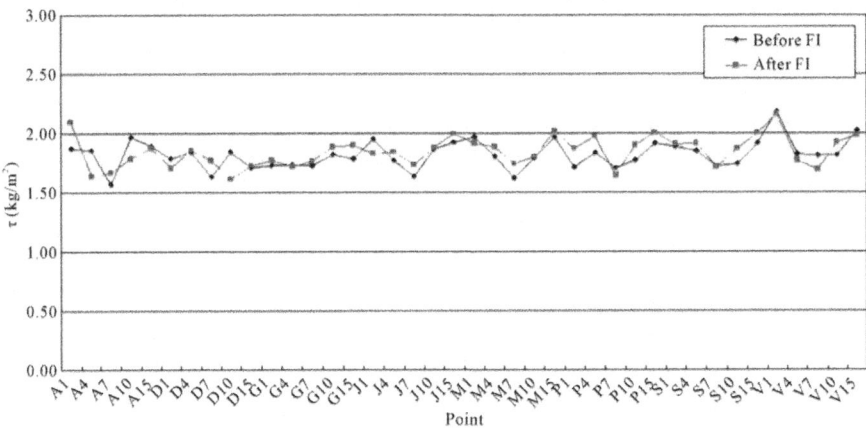

Figure 14: Comparison of shear stresses around the floating islands with frequency of 200 years.

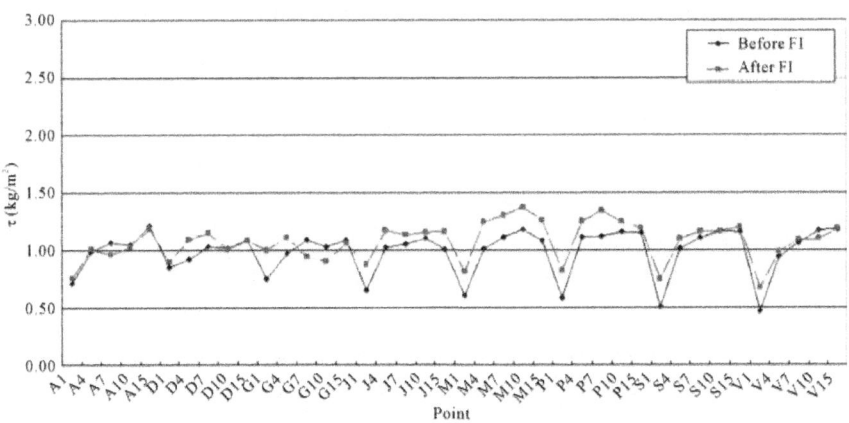

Figure 15: Comparison of shear stresses around the floating islands with bankfull discharge.

Floating Islands Layout

This study presented 2 revised proposals that can reduce whirlpool since the whirlpool was formed surrounding the floating islands in Figure 16. Figure 17 shows the flow duration surrounding the 2 revised proposals on floating islands. The left side of the floating islands installation section is the stagnant area with weak whirlpool before the floating islands installation, and solving with layout change alone is judged to be difficult. In the case of revised proposal 2, in the downstream area of the floating islands, the reduction of whirlpool's impact was confirmed because of smooth flow; thus, good flow is deemed to be formed compared with the design proposal and revised proposal 1.

Differences in bottom velocity and shear stress around the floating islands were found to be minimal in the design proposal (revised proposals 1 and 2; see Figure 18).

CONCLUSIONS

This study comparatively analyzed water levels, velocity distributions, and flow durations with regard to flood frequency of 200 years and riverside flood discharge (7066 m^3/s) in the Banpo zone to review the impacts of the floating island design proposal on the safety of flood control.

Although water level rise was locally observed due to the clash between the flow and the floating islands in terms of the surrounding water level change following floating islands installation, no water level rise impact was observed based on the water level comparison at the upstream measurement lines. There was no big difference by flood frequency in water level impacts due to the clash between the flow and the floating islands as a local phenomenon, which was observed within the section of about 5 - 10 m upstream of the floating islands. The experiment result suggests that the installation of the floated structure within the river does not have a huge effect on the water level of the relevant river section.

The analysis of the changes in bottom velocity and shear stress surrounding the floating islands was carried out for the purpose of predicting riverbed change that may be caused by the floating islands installation.

Although change of maximum shear stress exists, overall distribution difference was minimal according to the measurement points. The increase in shear stress arising from the floating islands installation was minimal; thus seemingly not affecting the riverbed. In the case of a large river with sufficient depth, the increase in shear stress owing to floating structure is not considered huge. When the structure is close to a riverbank, however, this can cause an increase in surface velocity under the riverside discharge condition, and this needs to be taken into account. Based on the experiment result of this study, a proposal to reduce the external force around the riverbank by changing flow duration was presented.

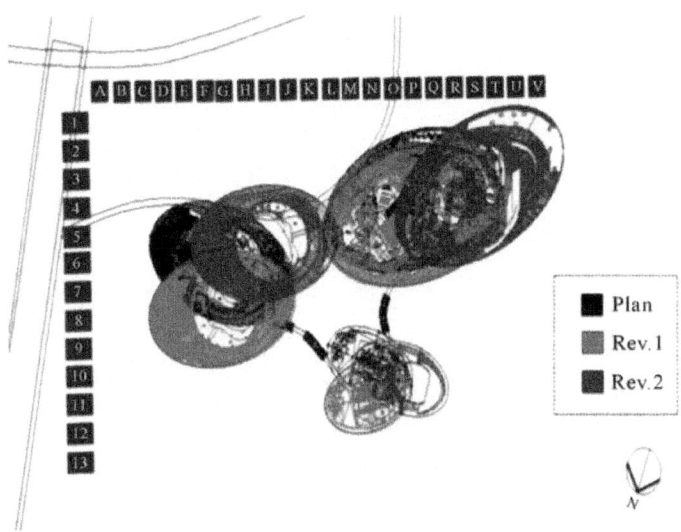

Figure 16: Revised proposal of layout of floating islands.

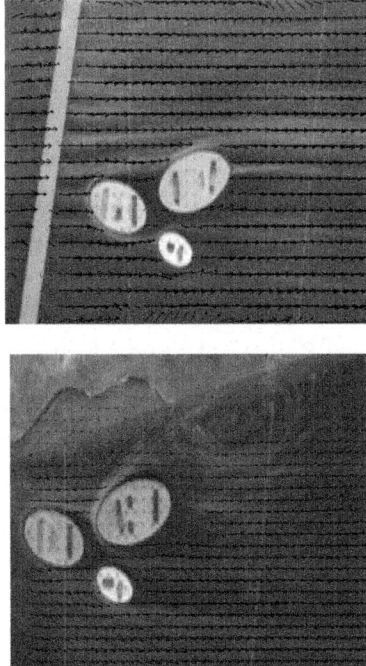

Figure 17: Planned cross sectional surface current field after floating islands (Frequency of 200 years and bankfull discharge).

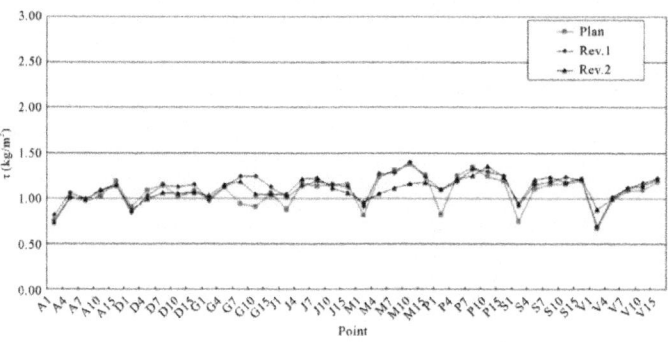

Figure 18: Comparison of shear stresses surrounding the floating islands.

REFERENCES

1.   A. Harpold and S. Mostaghimi, "Stream Discharge Measurement Using a Large-Scale Particle Image Velocimetry Prototype," Proceedings of Virginia Water Research Symposium, Blacksburg, 4-6 October 2004, pp. 178-185.

2.   ASCE Task Committee, "Hydraulic Modeling: Concepts and Practice," Manuals and Reports on Engineering Practice No. 97, ASCE, Reston, 2000.

3.   G. Dramaisa, J. L. Coza and B. C. A. Hauetb, "Advantages of a Mobile LSPIV Method for Measuring Flood Discharges and Improving Stage-Discharge Curves," Journal of Hydro-Environment Research, Vol. 5, No. 3, 2011, pp. 301-312. doi:10.1016/j.jher.2010.12.005

4.   R. J. Adrian, "Particle Imaging Techniques for Experimental Fluid Mechanics," Annual Review of Fluid Mechanics, Vol. 23, No. 1, 1991, pp. 261-304.doi:10.1146/annurev.fl.23.010191.001401

5.   M. Raffel, C. Willert and J. Kompenhans, "Particle Image Velocimetry: A Practical Guide," Springer, New York. 1998.

6.   E. A. Meshleh, T. Peeva and M. Muste, "Large Scale Particle Image Velocimetry for Low Velocity and Shallow Water Flows," Journal of Hydraulic Engineering, Vol. 130, No. 9, 2004, pp. 937-940. doi:10.1061/(ASCE)0733-9429(2004)130:9(937)

Chapter 4

LINKING POPULUS EUPHRATICA HYDRAULIC REDISTRIBUTION TO DIVERSITY ASSEMBLY IN THE ARID DESERT ZONE OF XINJIANG, CHINA

Xiao-Dong Yang[1, 2], Xue-Ni Zhang[3, 4], Guang-Hui Lv[3, 4], Arshad Ali[1, 2]

[1] College of Ecological and Environmental Sciences, East China Normal University, Shanghai, China

[2] Tiantong National Forest Ecosystem Observations and Research Station, Chinese National Ecosystem Observation and Research Network, Ningbo, China

[3] Institute of Resources and Environment Science, Xinjiang University, Urumqi, China

[4] Xinjiang Key Laboratory of Oasis Ecology, Ministry of Education, Urumqi, China

ABSTRACT

The hydraulic redistribution (HR) of deep-rooted plants significantly improves the survival of shallow-rooted shrubs and herbs in arid deserts, which subsequently maintain species diversity. This study was conducted in the Ebinur desert located in the western margin of the Gurbantonggut Desert. Isotope tracing, community investigation and comparison analysis were employed to validate the HR of *Populus euphratica* and to explore its effects on species richness and abundance. The results showed that, *P. euphratica* has HR. Shrubs and herbs that grew under the *P. euphratica* canopy (under community: UC) showed better growth than the ones growing outside (Outside community: OC), exhibiting significantly higher species richness and abundance in UC than OC ($p<0.05$) along the plant growing season. Species richness and abundance were significantly logarithmically correlated with the *P. euphratica* crown area in UC ($R^2=0.51$ and 0.84, $p<0.001$). In conclusion, *P. euphratica* HR significantly ameliorates the water conditions of the shallow soil, which then influences the diversity assembly in arid desert communities.

INTRODUCTION

The hydraulic redistribution (HR) is defined as the movement of water from the moist to dry soil portions through roots of deep-rooted plants [1]–[3]. Through this process, water moves upward or laterally among different soil layers, and subsequently changes the water spatial pattern of soil [3], [4]. Thus, HR improves the fine roots survival rate of deep-rooted plants and protects the shallow-rooted plants growth in extremely dry environments [4]–[6]. In addition HR plays an important role in positive feedback between plants and soil moisture circulation [1], [7]–[10], it also benefits the plant rhizosphere nutrient absorption, community stability, and biodiversity maintenance [3], [11]–[20].

Populus euphratica Oliv a deciduous plant with high tolerance to drought and saline conditions, and is a dominant species for the Tugai Forest (desert riparian forest) in the Ebinur desert. Pervious research findings have revealed that the water gradient of soil layers determines the occurrence of HR in deep-rooted plants [3], [4], [11], such as when the surface soils are extremely dry and the deeper soils are rich in available water [3], [4], [21]. P. euphratica developed a deep vertical root system for absorbing water in Chinese arid deserts due to effect of high wind [22]–[24]. Moreover, the annual precipitation of this site is less than 100 mm and the surface soils are extremely dry, whereas the groundwater level is relatively high because of glacial melt waters [22]. Hence, P. euphratica should possess HR.

In our previous study [22] and the one conducted by Hao et al. [23] employed the Ryel model [25] that was used to simulate HR of P. euphratica. According to the research findings of the above mentioned studies that P. euphratica has HR which reached a maximum point at 2:30 am for a given day, and the total amount of HR decreased with changes in the plant growing season [22], [23]. But, those results were only based on water characteristics (water content and water potential) changes along the various soil layers, which are only obtained from model predictions. The experimental support of P. euphratica possessing HR is lacking. Here we investigate whether oxygen isotope tracking can be used to test whether the plants possess HR [1]–[4], [11], and this could provide the experimental support to the studies based on model simulations. As oxygen isotope does not fractionate during water transport through xylem vessels, whereas the $^{18}O/^{16}O$ values varied across the different soil layers [1], [26]–[28]. For example, the $\delta^{18}O$ content continues to increase from deep to shallow soils based on fractionation. Thus, we can predict that if P. euphratica possess HR, P. euphratica roots can lift water containing lower $\delta^{18}O$ content from groundwater and deep soils to shallow soils through xylem vessels, which can result in soil having lower $\delta^{18}O$ content under the P. euphratica canopy (under community: UC) rather than outside the P. euphratica canopy (outside community: OC). Annual precipitation in Ebinur desert is very scarce, which is insufficient for the basic physiological activity of plants, while the deep soils and shallow groundwater contains abundant water for plant growth in this region [29]. However root density of most plants decreases with increasing soil depth [22], [30], [31], and only a small number of deep-rooted plants are able to penetrate to the deeper soils or the water table to absorb water which limits water utilization and subsequently, decreases the species richness and abundance in the arid desert [32], [33]. So, if P. euphratica HR lifts water from the deeper soils or groundwater aquifer to shallow soils through its roots, this process can increase the water content of surface soils and afford more water for shallow-rooted shrubs and herbs. In this case, it can induce the plants assembly under P. euphratica canopy and subsequently maintain the biodiversity in arid deserts. Therefore, we hypothesize that, the shallow-rooted shrubs and herbs grow better in UC than in OC and there is a higher species richness and abundance in UC than in OC, because P. euphratica HR significantly improves the soil water content. But, the correlation between P. euphratica HR, plant growth condition and biodiversity in arid desert areas is poorly understood.

The objective of this study was to test for P. euphratica HR and to explain the effect of HR on plants growth condition and species biodiversity maintenance in the Ebinur desert. We predict the following: (1) $\delta^{18}O$ content of the shallow soils in UC are higher than in OC; (2) shrubs and herbs of UC have higher growth condition than those of OC; and (3) P. euphratica HR influence species abundance and richness in arid desert.

MATERIALS AND METHODS

Study site and Ethics Statement

The experimental site is located in the Ebinur Lake Wetland Nature Reserve (ELWNR) in the western margin of the Gurbantonggut Desert in Xinjiang Uygur Autonomous Region of China (44° 30'–45° 09' N, 82° 36'–83° 50' E). This site belongs to a tuyere zone of Alashankou, the annual wind days (days with wind speed $\geqq 17$ m/s) are more than 164 d, and the annual fresh gale hours are approximately 241 hr. The annual sunshine hours reach approximately 2800 hr, and the annual precipitation is less than 100 mm, whereas the potential evaporation is more than 1600 mm. In addition, the temperature in this area ranges from 44 to −33°C, with an average temperature ranging from 6 to 8°C, and an average temperature of the growing season is roughly 25°C. Due to the extremely dry conditions and sparse rainfall, the climate is classified as typical temperate continental arid [22]. No specific permissions were required for the described field studies in ELWNR. The ELWNR is owned and managed by the local government and the location including the site used for our experiment are not privately owned or protected in any way and thus a specific permit for not for-profit research is not required. The field studies did not involve endangered or protected plant species in this area.

Study Plot and samples Collection

5×5 km (25 km²) typical plots of P. euphratica were chosen as our experimental plots, where 48P. euphratica individuals are distributed and the vegetal coverage is 4%. This site has a moderate slope and sandy soil. The distribution of groundwater level in our experimental site ranges from 1.5 to 1.8 m. Moreover, sand dunes are present 3 km north of the site. Based on the terrain of the experimental site, we randomly selected three P. euphratica individuals with no differences in growth conditions (DBH, tree height and crown area) among those trees as the experimental (UC) group. The selected plant individuals were at least 6 m apart from each other to prevent mutual water transfer process among them. For comparison, we randomly selected three plots located closely to the selected three *P. euphratica* individuals as the control (OC) group, where no *P. euphratica* were growing and no groundwater table difference with UC group.

To distinguish the difference in water transport from deeper soil layers to surface soil layers between the UC and OC groups, three points in each two types of plots were randomly selected and each point was divided into five soil layers (0 to 10 cm, 10 to 40 cm, 40 to 70 cm, 70 to 100 cm, and 100 to 150 cm, the deepest soil layer is 150 cm, as groundwater appeared approximately at 160 cm when the soil columns were dug out in the above selected OC and UC points). Each soil sample was dug with a soil auger between 4:30 AM and 5:30 AM (Xinjiang local time) in the middle of July, 2010. The nine soil samples for each soil layer from three points were mixed to makeup a composite soil samples. Meanwhile, on the periphery of selected plots, well water 3 m beneath ground level was collected from five sites, and subsequently mixed them to obtain a composite sample to replace groundwater. A river water sample was also mixed from three sites at least 1 km apart of the Aqikesu River. Each sample of river water was collected from 1 m beneath river surface. All the soil and water samples were immediately placed into glass bottles, sealed with Parafilm and stored in a mini refrigerator (4°C) and brought to the Xinjiang University Physiological Ecology Laboratory for further experimental analysis.

Oxygen Isotope Measurement

Methods for extraction of soil water are consistent with Allison et al. [34]. Water isotope content ($\delta^{18}O$) was measured using a DELTA V Advantage Isotope Ratio Mass Spectrometer (Thermo, Waltham, MA, USA) at the Chinese Academy of Forestry Stable Isotope Laboratory. Each sample was measured three times continuously with the third result as the experimental oxygen isotope value. Precision values of continuous measurements for standard sample were as follows: D, <3‰ and ^{18}O, <0.5‰. The isotopic abundance was expressed in delta notation (δ) in parts per thousand (‰) as

$$\delta = (R_{sample}/R_{standard} - 1) \times 1,000 \qquad (1)$$

Where Rsample and Rstandard are the molar ratios of heavy to light isotope of the sample and the international standard (Vienna standard mean ocean water for $^{2}H/^{1}H$ and $^{18}O/^{16}O$) [1].

Plots Investigation

48 plots of UC were established having each plot size of 10×10 m (14 plots in June, 17 plots in August and October) in 5×5 km experimental plot from June to October, 2010. Consequently, 34 control plots with 10×10 m of OC (13 plots in June, 15 plots in August and 16 plots in October) were set at the same time. The species identification, richness, abundance, coverage, DBH, and crown area of each plant in each plot were investigated.

Data Analysis

To determine whether *P. euphratica* undergoes HR, a comparison analysis was used to show the difference in $\delta^{18}O$ content for the five soil layers between the UC and OC groups. In addition, the paired-sample t-test was used to test the significant difference between above mentioned two groups.

Furthermore, based on the plots investigation data, the growth dominance index (Formulae 2 to 4) was used to show the growth condition, i.e. difference between UC and OC.

$$x_s = 1 - \frac{x_{max} - x}{x_{max} - x_{min}} \qquad (2)$$

$$DI = \frac{x_{sc} + x_{sa} + x_{sh} + x_{sf}}{400} \qquad (3)$$

$$GDI = \frac{\sum_{i=1}^{N} DI_i}{N} \qquad (4)$$

Where x_s, x_{sc}, x_{sa}, x_{sh} and x_{sf} are the standardized value, coverage, abundance, height and frequency for a given species in a given plot respectively. x_{max} and x_{min} are the maximum and minimum values for a given species in a given plot while x is the actual investigative value of each individual belonging to the same species in a given plot. DI is the dominance index in a given plot, and GDI is the growth dominance index of all plots. N is the count of plots, and DI_i is the dominance index of a given species in i th plot.

Individual size has positive correlations with the amount of HR which can generally represent the water lift capacity of HR. It is well understood that the crown area of arid desert vegetation is significantly linked with adult size (tree height) and root biomass [35], [36]. Thus in this study, we used the individual crown area (CA, Formula 5) to represent water lifting capacity of *P. euphratica* HR. Maximum vegetation crown diameter (CD1) and its perpendicular diameter (CD2) were measured on each plant individual, which were used to calculate CA as follows (Formula 5).

$$CA(m^2) = CD1(m) \times CD2(m) \times \frac{\pi}{4}$$

(5)

In order to explain the effects of HR on species richness and abundance in the arid desert community, an independent sample t-test was used between UC and OC to determine the diversity difference for shrubs and herbs. Finally, the logarithmic regression analysis was applied to explore the relationship of HR capacity (CA) to species richness (species number in 10×10 m) and abundance (count of individual plants in 10×10 m) for arid desert community.

All statistical tests were conducted using SPSS 11.5 while related figures were drawn using Origin 8.0. All statistical tests were considered significant at the $p < 0.05$ level.

RESULTS

Comparisons of $\delta^{18}O$ values between the UC and OC

$\delta^{18}O$ content of the OC were generally higher than that of the UC among five soil layers. In addition, the maximum difference between the two groups were observed in the surface soil layer (0 to 10 cm), and then decreased with soil depth (Fig. 1a). Therefore, a paired-sample t-test exhibited that the mean of five soil layer $\delta^{18}O$ contents of OC was significantly higher than that of UC ($t = -3.28$, $df = 4$, $p < 0.05$, Fig. 1b). The general comparison showed that $\delta^{18}O$ content was the highest in soil, medium in groundwater and the lowest in river (Fig. 1a).

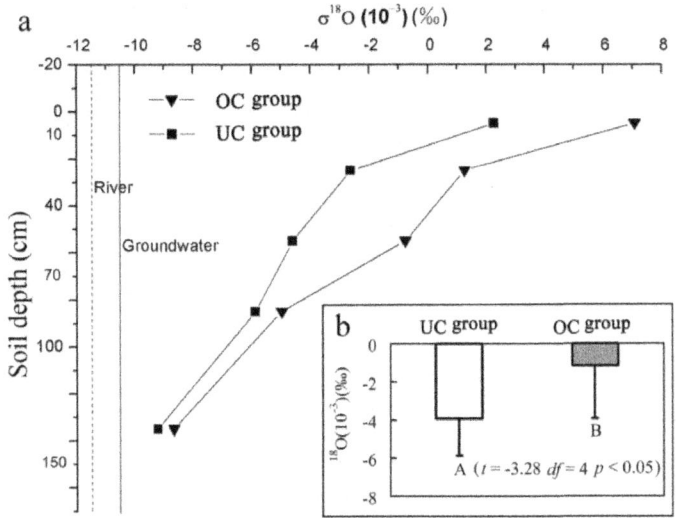

Figure 1. Difference in δ¹⁸O value with soil depth between UC and OC groups.

UC is the community under the *P. euphratica* canopy, whereas the OC is the community outside the *P. euphratica* canopy. The soils were divided into five layers (0 to 10 cm, 10 to 40 cm, 40 to 70 cm, 70 to 100 cm, and 100 to 150 cm) based on *P. euphratica* roots and underground water level distributions. Each point in the Fig. 1a showing δ¹⁸O for each soil layer sample, which was measured for three times continuously with the third result as the experimental oxygen isotope value. Nine soil samples for each soil layer from three points were mixed to get one composite soil in each group. ¹⁸O contents of river (−11.55‰) and underground water (−10.59‰) are show through vertical dashed and solid lines respectively in Fig. 1a. Fig. 1b shows the mean comparison between UC and OC groups and the data in parenthesis showing the Paired-Sample *t*-test result. (Mean ± *SD*) for significance difference between two groups.

doi:10.1371/journal.pone.0109071.g001

Difference in Species Dominance Index and Biodiversity between the UC and OC

Both the shrubs and herbs growth dominance index (*GDI*) of UC were higher than that of OC in June, August and October (Table 1). This finding indicates that the shrubs and herbs species grew better in UC than that in OC. Furthermore, compared with OC, the shrubs richness and abundance of the UC were significantly higher across three months (*p*<0.05) (Fig. 2). But for the herbs, this pattern of richness and abundance varies along the plant growing season. Between UC and OC, the species richness and abundance of herbs and shrubs were significantly similar in June and August, while non-significant differences were found in October (Fig. 2). Further, based on all UC plots investigated, the logarithmic regression analysis was used to analyze the relationship of *P. euphratica* individual HR capacities (individual crown area) with species richness and abundance. The results show that, species richness and abundance exhibited a significant exponent correlation with *P. euphratica* CA ($p<0.05$) (Fig. 3).

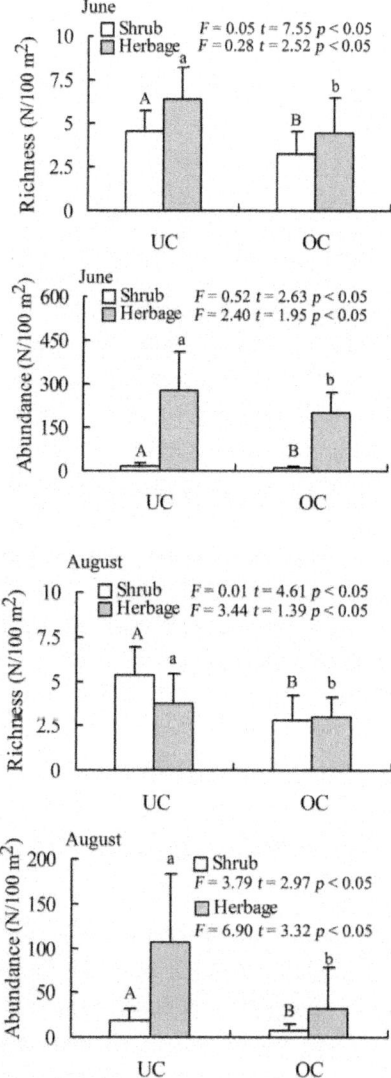

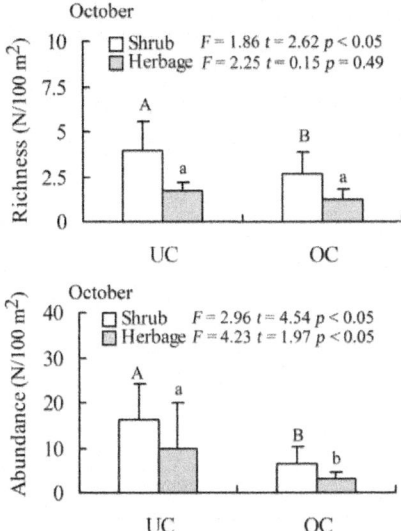

Figure 2. Difference in richness and abundance along plant growing season between UC and OC. UC is the community under the *P. euphratica* canopy, whereas the OC is the community outside the *P. euphratica* canopy. Blank and grid boxes indicate shrub and herbage, respectively. The Independent sample *t*-test is used to analyze the differences of richness and abundance between UC and OC. Different capital letters on each blank box indicate significant differences of shrub richness or abundance between UC and OC. Different lowercase letters on each grid box indicate significant differences of herbage richness or abundance between UC and OC. *p*<0.05. Numbers in figure are the results of Independent sample *t*-test. (Mean ± *SD*) for significance difference between UC and OC.

doi:10.1371/journal.pone.0109071.g002

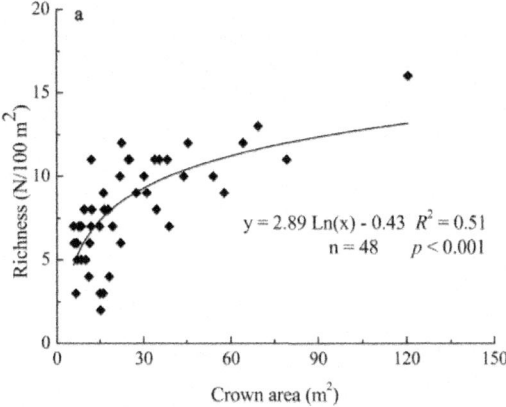

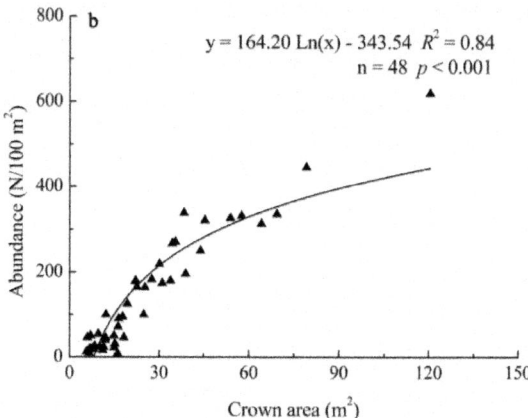

Figure 3: Logarithmic regression of *P. euphratica* crown areas against species abundance and richness. In regression line (Fig. 3a), each point indicating the richness of all species (no including *P. euphratica*) under the *P. euphratica* canopy. In regression line (Fig. 3b), each point indicating the abundance of all species (no including *P. euphratica*) under the *P. euphratica* canopy. The crown area of each sample is the sum of *P. euphratica* individuals. doi:10.1371/journal.pone.0109071.g003

Table 1: Differences of species growth dominance index between the UC and the OC.

Life form	Species	Growth dominance index					
		June		August		October	
		UC	OC	UC	OC	UC	OC
Shrub	Kalidium foliatum	0.29	0.27	0.31	0.19	0.19	0.12
	Reaumuria soogorica	0.54	0.35	0.53	0.33	0.41	0.41
	Alhagi sparsifolia	0.67	0.50	0.29	0.22	0.34	0.25
	Lycium ruthenicum	0.23	0.21	0.35	0.32	0.26	0.09
	Halostachys caspica	-	0.06	0.26	0.33	0.16	0.21
	Calligonum L	0.15	0.08	0.05	-	0.05	0.05
	Populus euphratica	0.41	0.18	0.00	-	0.05	-
	Nitraria schoberi	0.20	0.13	0.12	-	-	0.05
	Haloxylon persicum	0.22	-	0.31	0.05	0.32	0.21
	Salsola passerina	0.05	-	0.20	0.03	0.15	0.05
	Nitraria sibirica			0.15	0.10	0.12	0.12
	Tamarix ramosissina	-	-	0.14	-	0.13	-
Herbage	Karelinia caspicos	0.42	0.25	0.14	0.11	0.05	0.05
	Salsola soda	0.67	0.51	0.54	0.45		
	Salsola nitraria	0.31	0.31	0.30	0.43	0.18	0.14
	Poacynum hendersoni	0.55	0.44	0.53	0.41	-	-
	Spriphidium	0.34	0.11	0.09	-	0.05	-
	Suaeda glauca	-	-	0.22	0.10	0.17	0.05
	Aeluropus littoralis	0.07	0.06	0.05	-	0.05	-
	Phragmites australis	-	-	0.17	0.30	0.15	0.05
	Scorzonera austriaca	0.11	0.18	0.10	-	-	-
	Carpesium abrotonoides	0.20	0.06	-	-	-	-
	Atriplex patens	0.08	0.06	-	-	-	-
	Petrosimonia sibirica	0.09	-	-	-	-	-
	Malcolmia africana	0.38	0.12	-	-	-	-

UC is the community under the *P. euphratica* canopy, whereas the OC is the community outside the *P. euphratica* canopy. For UC, *P. euphratica* is the only tree species (woody plant, height >6 m) while the others are shrub (woody plant, height <6 m) and herbage (herbaceous plants, height <1 m). For OC, all plants are shrubs and herbages. "-" in table indicating no values because of the ephemeral plants life history turnoff and randomly setting samples along June (early growth period of *P. euphratica*), August (middle growth period of *P. euphratica*) and October (defoliating period of *P. euphratica*). The figures in table showing Growth dominance indices of species belonged to UC and OC. All plots are established in 5×5 km area of Ebinur Lake Wetland Nature Reserve in Xinjiang Uygur Autonomous Region of China.
doi:10.1371/journal.pone.0109071.t001

DISCUSSION

Experimental Validation of *P. euphratica* HR

Oxygen isotope fractionates during water transport between the different soil layers because of physical and chemical adsorption and conduction processes, which changes the $^{18}O/^{16}O$ values in different soil layers. However, the fractionation of oxygen isotopes does not occur during water transport through the xylem vessels [1], [26]–[28]. So the water originating from the deeper soils or groundwater in the shallow soils is an indicator for HR occurrence [1]–[3],[11], [37]. In the present study, the soil samples of UC and OC groups were collected at between 4:30 AM and 5:30 AM (local time), during which the fractionation not occur due to soil evaporation. Therefore, if *P. euphratica* HR does not occur, as the environmental conditions were same between UC and OC groups, the $\delta^{18}O$ content of the corresponding soil layers did not differ significantly between UC and OC groups. However, in this study, all $\delta^{18}O$ content of UC were lower than those of OC group among five soil layers (Fig. 1a), and the mean of UC was also significantly lower than OC group (Fig. 1b). This overall pattern exhibited that the water of all UC soil layers partially originated from deeper soil or groundwater through plants vessel transport. In other words, the deep-rooted *P. euphratica* exhibited HR, which transported and lifted water from deeper to shallow soil layers, proof for the experimental support to Ryel model based studies [22].

It is well understood that changes in soil evaporation depends on vegetation coverage, for example, the higher vegetation coverage will has more shade on the soil surface and hence evaporation will decrease, and vice versa [38], [39]. Furthermore, Allison [40] and Kim [41] confirmed that $\delta^{18}O$ content of soil samples in UC is lower as compared to OC, because high vegetation coverage decreased the evaporation due to shading. Specifically, higher evaporative demand in OC could easily drive a greater upward movement of water from the groundwater or the deeper soils to surface soils, which then enriched the $\delta^{18}O$ content due to evaporative fractionation. However, pan evaporation was not significantly different between UC and OC ($UC_{soil\ evaporation}=0.91\pm0.52$ $cm \cdot m^{-2} \cdot d^{-1}$, $OC_{soil\ evaporation}=0.93\pm0.55$ $cm \cdot m^{-2} \cdot d^{-1}$, $t=-1.02, df=6, p=0.34$, which was tested through paired-sample t-test, and its supportive data was measured by 255 Series Evaporation Stations (EP255, Novalynx Inc, OR, USA) for three paired groups in UC and OC), because the sparse vegetation coverage of *P. euphratica* community (approximate 4%) has little shading influence on the evaporative demand of UC and OC in Ebinur desert [22]. In addition, this desert is a part of tuyere zone having high wind, and thus other evaporative environmental factors (*e.g.*, air temperature and moisture) homogenously influence soil evaporation, subsequently caused relatively no difference of soil evaporation between UC and OC plots. Hence, soil evaporation not likely differed $\delta^{18}O$ content between OC and UC in the Ebinur desert and further proved that *P. euphratica* HR occurs.

P. euphratica HR and water transport direction along soil profile can also be judged by moisture differences among soil layers. In this study, Thetaprobe ML2 soil moisture sensors (Delta-T Devices, Cambridge, UK) were installed respectively in five soil layers of UC and OC to measure the variation in soil volumetric water content (SVWC) along soil depth. The results showed that, (1) SVWCs of 0 to 10 cm, 10 to 40 cm, 70 to 100 cm and 100 to 150 cm layers in UC were significantly higher than those in OC, except in 40 to 70 cm soil layer; (2) SVWC increases with soil depth but decreases with $\delta^{18}O$ content in OC, while no significant trend was found in UC; and (3) SVWC of 100 to 150 cm soil layer was a little higher in OC than in OC (Table 2). These results suggested that groundwater supplies soil water, and the water's table in both UC and OC were similar in Ebinur desert. SVWC increased with soil depth due to the extraction of soil evaporation in OC. But HR can lift water from groundwater and deep soils to shallow soils and then decrease the effect of evaporation on water extracting in UC. Previous studies showed that the location and the amount of HR releasing water were significantly depended on plant fine roots distribution [11], [22], [23]. Also *P. euphratica*'s fine roots was mainly growing within 0 to 70 cm soil depth in arid area [22], [29]. Thus, HR may cause no change in SVWC across soil depth in UC and SVWC of 40 to 70 cm layer in UC less than that in OC.

Table 2: Variation in soil volumetric water content across soil depth and $\delta^{18}O$ content in UC and OC.

Soil layers (cm)	$\delta^{18}O$ (‰)		Soil volumetric water contents (Mean ± SD) (cm³/cm³)		F	t	P
	UC	OC	UC	OC			
0–10	2.27	7.10	8.41±0.38A	3.86±1.67B	45.60	21.79	<0.001
10–40	−2.62	1.25	11.71±0.11A	18.86±0.26B	1.95	181.58	<0.001
40–70	−4.60	−0.77	12.63±0.45A	20.26±0.35B	16.03	−100.46	<0.001
70–100	−5.87	−4.98	23.07±0.40A	21.40±0.07B	2.08	34.07	<0.001
100–150	−9.21	−8.67	27.98±0.03A	27.55±0.02B	66.78	7.49	<0.05

UC is the community under the *P. euphratica* canopy, whereas the OC is the community outside the *P. euphratica* canopy. UC and OC have three measuring plots, respectively. The independent sample t-test is used to analyze the differences of soil volumetric water content between UC and OC. Different capital letters in each row indicate significant differences of soil volumetric water content between UC and OC. The measurement period of soil volumetric water content lasts from 3th to 8th August, 2010. During the time between 2:30 am to 6:30 am, 11:30 am to 5:00 pm at each experimental day, the soil volumetric water content recorded manually at intervals of 2 hours, while the time between 6:30 am to 9:30 am, 9:30 pm to 12:30 am, the soil volumetric water content recorded manually once an hour.
doi:10.1371/journal.pone.0109071.t002

The greatest $\delta^{18}O$ content existed in soil while the intermediate in groundwater and the lowest in river (Fig. 1a). This pattern indicates that river is the initial source of water for arid desert area. Our results were supported by Zhao et al. [42] reported that river largely originated from glacial melting and has the lowest $\delta^{18}O$ content in arid desert. Based on oxygen isotope fractionation theory [26]–[28], [40], initial water resource has lowest $\delta^{18}O$ content and then increased with transmittal distance and pathways. In this study, the river is the main source for groundwater and then supply to soil water through water transmittal ways i.e. underground or surface runoff, evaporation and HR. All of those can led to oxygen isotope fractionation and then resulted in soil having the greatest value of $\delta^{18}O$ content.

Effects of *P. euphratica* HR on species growth condition

The importance value index [importance value index=(relative abundance + relative frequency + relative coverage + relative height)/400] and its deduced dominance index (dominance index=$\sum_{i=1}^{n}\left(\frac{n_i}{N_i}\right)^2$, N is the count of plots and N_i is the importance value of a certain species in i th plot) were traditional ecological methods in generally using to evaluate species growth conditions within a specific community type [43], [44]. Specifically, the importance value is commonly assumed as 1, and then based on individual relative statistical values, such as individual numbers, relative abundance, relative coverage, relative frequency, and relative height. 1 was divided into different components values that show the different species growth conditions within a given community. Nevertheless, many studies also needed to compare the difference of same species growth conditions among environmental sites, such as the same species between UC and OC. According to this situation, considering *P. euphratica* can account for the largest partition of the importance value in UC, if we used the traditional methods above to evaluate the difference of species growth conditions between UC and OC, a lower importance value could be found in UC than in OC among other species, even these species have the same abundance, coverage, frequency and height between two environmental types.

In this study, the growth dominance index was structured by the actual values of coverage, abundance, height, and frequency of a given species in a given plot to avoid the components influence of importance value, and to compare the difference of species growth conditions between the UC and OC (Formulae 2 to 4). The results showed that UC species grow better than OC along plant growing seasons (Table 1). Additionally SVWC increased with soil depth in OC while no change in UC (Table 2). These suggest that *P. euphratica* HR increased significantly water content of shallow soils and benefited the growth condition and survival of shallow-rooted plants [16], [17], [24], [37], [45]–[50].

Relationship between *P. euphratica* HR and arid Desert Diversity Assembly

Water is the main limitation for arid desert plant communities [1], [51]. Thus, species coexistence pattern in this community directly depends on spatial availability and distribution of water [32], [33]. In this study, the species richness and abundance in the UC were significantly higher than those in the OC ($p<0.05$), i.e., UC has higher species diversity (Fig. 2).

Furthermore, based on the hypothesis of this study that a tree with a larger crown area has higher root biomass and transports more water from the deep to shallow soils, the logarithmic regression was used to analyze the relationship between the *P. euphratica* HR capacity (CA) and the species richness and abundance in the UC. The results showed that the *P. euphratica* CA was significantly logarithmically correlated with richness ($R^2=0.51$) and abundance ($R^2=0.84$) ($p<0.001$) (Fig. 3). These results indicated that species diversity was influenced significantly by HR. Similarly, this conclusion was also reflected in other studies. For example, herbaceous plants under tree canopies have higher abundance and richness [46], [52], and the presence of trees promoted the survival of shrubs and seedlings grown under tree canopies [37], [45], [48]–[50], [53].

Shallow-rooted herbs and shrubs that absorb water from *P. euphratica* HR provided more organic matter and mineral elements to the soil of the UC than to those of the OC because of the "fertile island" effect. It implies that more individuals of the UC are intercepted and sequester larger amounts of organic matter and minerals from surface winds because of increased total crown area [54]–[59]. Meanwhile, the more numerous individuals also produced more litter into the UC soils than OC soils. Thus, these processes contributed to the nutrient absorption and growth of *P. euphratica*, which further increased its HR capacity. In turn, these processes also increased the shading in understory plants, which can decrease the transpiration of herbs and shrubs, and further improve its survival. Therefore, there appears to be a species coexistence pattern of water sharing and resource complementation between the deep-rooted *P. euphratica* and other shallow-rooted species, as well as positive feedback between *P. euphratica* HR and biodiversity maintenance in arid deserts community.

Acknowledgments

The authors thank Yang jun, Sun Jingxing, Tian Youhua, Sun Lijun, Zhang Xuemei, Li Changjun, He Xueming, Qin Lu, Luo Chong, Ren Manli, Zhu Ya for their help in the field and laboratory. The authors also thank Qao Qiang and Dearlyn Fernandes for English grammar checking of the manuscript.

Author Contributions

No other contributions. Conceived and designed the experiments: XDY GHL.

Performed the experiments: XDY XNZ. Analyzed the data: XDY. Contributed reagents/materials/analysis tools: XDY GHL. Wrote the paper: XDY AA.

REFERENCES

1. Armas C, Padilla F, Pugnaire F, Jackson R (2010) Hydraulic lift and tolerance to salinity of semiarid species: consequences for species interactions. Oecologia 162: 11–21.

2. Haase P, Pugnaire FI, Ferna´ndez EM, Puigdefa´bregas J, Clark S, et al. (1996) An investigation of rooting depth of the semiarid shrub Retama sphaerocarpa (L.) Boiss by labelling of ground water with a chemical tracer. Journal of Hydrology 177: 23–31.

3. Richards JH, Caldwell MM (1987) Hydraulic lift: substantial nocturnal water transport between soil layers by Artemisia tridentata roots. Oecologia 73: 486– 489.

4. Caldwell MM, Dawson TE, Richards JH (1998) Hydraulic lift: consequences of water efflux from the roots of plants. Oecologia 113: 151–161.

5. Hacker SD, Bertness MD (1995) Morphological and physiological consequences of a positive plant interaction. Ecology 76: 2165–2175.

6. Vetterlein D, Marschner H (1993) Use of a microtensiometer technique to study hydraulic lift in a sandy soil planted with pearl millet (Pennisetum americanum [L.] Leeke). Plant and Soil 149: 275–282.

7. Filella I, Penuelas J (2003) Indications of hydraulic lift by Pinus halepensis and its effects on the water relations of neighbour shrubs. Biologia Plantarum 47: 209– 214.

8. Franco A, Nobel P (1990) Influences of root distribution and growth on predicted water uptake and interspecific competition. Oecologia 82: 151–157.

9. Kurz-Besson C, Otieno D, do Vale RL, Siegwolf R, Schmidt M, et al. (2006) Hydraulic lift in cork oak trees in a savannah-type Mediterranean ecosystem and its contribution to the local water balance. Plant and Soil 282: 361–378.

10. Williams DG, Ehleringer JR (2000) Intra-and interspecific variation for summer precipitation use in pinyon-juniper woodlands. Ecological Monographs 70: 517– 537.

11. Burgess SS, Adams MA, Turner NC, Ong CK (1998) the redistribution of soil water by tree root systems. Oecologia 115: 306–311.

12. Dawson TE (1993) Hydraulic lift and water use by plants: implications

for water balance, performance and plant-plant interactions. Oecologia 95: 565–574.

13. Lee J-E, Oliveira RS, Dawson TE, Fung I (2005) Root functioning modifies seasonal climate. Proceedings of the National Academy of Sciences of the United States of America 102: 17576–17581.

14. Lehto T, Zwiazek JJ (2011) Ectomycorrhizas and water relations of trees: a review. Mycorrhiza 21: 71–90.

15. Neumann RB, Cardon ZG (2012) the magnitude of hydraulic redistribution by plant roots: a review and synthesis of empirical and modeling studies. New Phytologist 194: 337–352.

16. Prieto I, Martı´nez-Tilleri´a K, Martı´nez-Manchego L, Montecinos S, Pugnaire FI, et al. (2010) Hydraulic lift through transpiration suppression in shrubs from two arid ecosystems: patterns and control mechanisms. Oecologia 163: 855–865.

17. Warren JM, Brooks JR, Dragila MI, Meinzer FC (2011) In situ separation of root hydraulic redistribution of soil water from liquid and vapor transport. Oecologia 166: 899–911.

18. Zou C, Barnes P, Archer S, McMurtry C (2005) Soil moisture redistribution as a mechanism of facilitation in savanna tree–shrub clusters. Oecologia 145: 32–40.

19. David TS, Pinto CA, Nadezhdina N, Kurz-Besson C, Henriques MO, et al. (2013) Root functioning, tree water use and hydraulic redistribution in Quercus suber trees: A modeling approach based on root sap flow. Forest Ecology and Management 307: 136–146.

20. Sardans J, Pen~uelas J (2014) Hydraulic redistribution by plants and nutrient stoichiometry: Shifts under global change. Ecohydrology 7: 1–20.

21. Horton JL, Hart SC (1998) Hydraulic lift: a potentially important ecosystem process. Trends in Ecology & Evolution 13: 232–235.

22. Yang XD, Lv GH (2011) Establishment and analysis of Populus euphratica'root Hydraulic redistribution model. Chinese journal of Plant Ecology 35: 816–824 (in Chinese with English abstract).

23. Hao XM, Li WH, Guo B, Ma JX (2013) Simulation of the effect of root distribution on hydraulic redistribution in a desert riparian forest. Ecological research 28: 653–662.

24. Yu T, Feng Q, Si J, Xi H, Li Z, et al. (2013) Hydraulic redistribution of soil water by roots of two desert riparian phreatophytes in northwest China's extremely arid region. Plant and soil 372: 297–308.

25. Ryel R, Caldwell M, Yoder C, Or D, Leffler A (2002) Hydraulic redistribution in a stand of Artemisia tridentata: evaluation of benefits to transpiration assessed with a simulation model. Oecologia 130: 173–184.

26. Dawson TE, Mambelli S, Plamboeck AH, Templer PH, Tu KP (2002) Stable isotopes in plant ecology. Annual review of ecology and systematics 33: 507–559.

27. Ehleringer J, Dawson T (1992) Water uptake by plants: perspectives from stable isotope composition. Plant, Cell & Environment 15: 1073–1082.

28. Ellsworth PZ, Williams DG (2007) Hydrogen isotope fractionation during water uptake by woody xerophytes. Plant and Soil 291: 93–107.

29. Zhao F, Jin HL (2011) Study on characteristics of groundwater and its impact on Populus euphratica along the banks of Aqikesu River. Journal of Arid Land Resources and Environment 8: 156–161 (in Chinese with English abstract).

30. Hamblin AP (1985) the influence of soil structure on water movement, crop root growth and water uptake. Advances in Agronomy 38: 95–158.

31. Jackson R, Canadell J, Ehleringer J, Mooney H, Sala O, et al. (1996) A global analysis of root distributions for terrestrial biomes. Oecologia 108: 389–411.

32. Li XR, Tan HJ, He MZ, Wang XP, Li XJ (2009) Patterns of shrub species richness and abundance in relation to environmental factors on the Alxa Plateau: Prerequisites for conserving shrub diversity in extreme arid desert regions. Science in China Series D: Earth Sciences 52: 669–680.

33. Xia Y, Moore DI, Collins SL, Muldavin EH (2010) aboveground production and species richness of annuals in Chihuahuan Desert grassland and shrubland plant communities. Journal of arid environments 74: 378–385.

34. Allison GB, Hughes MW (1983) the use of natural tracers as indicators of soilwater movement in a temperate semi-arid region. Journal of Hydrology 60: 157– 173.

35. King DA (1996) Allometry and life history of tropical trees. Journal of tropical ecology 12: 25–44.

36. Poorter L, Bongers F, Sterck FJ, Wo"ll H (2003) Architecture of 53 rain forest tree species differing in adult stature and shade tolerance. Ecology 84: 602–608.

37. Callaway RM (1992) Effect of shrubs on recruitment of Quercus douglasii and Quercus lobata in California. Ecology 73: 2118–2128.

38. Monteith JL (1965) Evaporation and environment. Symposia of the

Society for Experimental Biology 19: 205–234.

39. Raz-Yaseef N, Rotenberg E, Yakir D (2010) Effects of spatial variations in soil evaporation caused by tree shading on water flux partitioning in a semi-arid pine forest. Agricultural and Forest Meteorology 150: 454–462.

40. Allison GB, Barnes CJ, Hughes MW (1983) the distribution of deuterium and 18O in dry soils 2. Experimental. Journal of Hydrology 64: 377–397.

41. Kim K, Lee X (2011) isotopic enrichment of liquid water during evaporation from water surfaces. Journal of hydrology 399: 364–375.

42. Zhao L, Xiao H, Cheng G, Song Y, Zhao L, et al. (2008) A preliminary study of water sources of riparian plants in the lower reaches of the Heihe basin. Acta Geoscientica Sinica 29: 709–718.

43. Zhang JT (2004) Quantitive Ecology. Beijing: Science Press (in Chinese).

44. Numata M (1979) Methods for Ecological Study. Kokin Shou-in, Tokyo (in Japanese).

45. Barchuk A, Valiente-Banuet A, Di′az M (2005) Effect of shrubs and seasonal variability of rainfall on the establishment of Aspidosperma quebracho-blanco in two edaphically contrasting environments. Austral ecology 30: 695–705.

46. Belsky A, Amundson R, Duxbury J, Riha S, Ali A, et al. (1989) The effects of trees on their physical, chemical and biological environments in a semi-arid savanna in Kenya. Journal of Applied Ecology 26: 1005–1024.

47. Belsky AJ (1994) Influences of trees on savanna productivity: tests of shade, nutrients, and tree-grass competition. Ecology 75: 922–932.

48. Franco A, Nobel P (1989) Effect of nurse plants on the microhabitat and growth of cacti. The Journal of Ecology 77: 870–886.

49. Maestre FT, Bautista S, Cortina J (2003) Positive, negative, and net effects in grass-shrub interactions in Mediterranean semiarid grasslands. Ecology 84: 3186–3197.

50. Shumway SW (2000) Facilitative effects of a sand dune shrub on species growing beneath the shrub canopy. Oecologia 124: 138–148.

51. Huxman TE, Snyder KA, Tissue D, Leffler AJ, Ogle K, et al. (2004) Precipitation pulses and carbon fluxes in semiarid and arid ecosystems. Oecologia 141: 254–268.

52. Weltzin JF, Coughenour MB (1990) Savanna tree influence on understory vegetation and soil nutrients in northwestern Kenya. Journal of Vegetation Science 1: 325–334.

53. Rousset O, Lepart J (1999) Shrub facilitation of Quercus humilis

regeneration in succession on calcareous grasslands. Journal of Vegetation Science 10: 493–502.

54. Armas C, Pugnaire FI (2005) Plant interactions govern population dynamics in a semi-arid plant community. Journal of Ecology 93: 978–989.

55. Jackson L, Strauss R, Firestone M, Bartolome J (1990) Influence of tree canopies on grassland productivity and nitrogen dynamics in deciduous oak savanna. Agriculture, Ecosystems & Environment 32: 89–105.

56. Reynolds JF, Virginia RA, Kemp PR, de Soyza AG, Tremmel DC (1999) Impact of drought on desert shrubs: effects of seasonality and degree of resource island development. Ecological Monographs 69: 69–106.

57. Schade JD, Hobbie SE (2005) Spatial and temporal variation in islands of fertility in the Sonoran Desert. Biogeochemistry 73: 541–553.

58. Segoli M, Ungar ED, Shachak M (2012) Fine-Scale Spatial Heterogeneity of Resource Modulation in Semi-Arid "Islands of Fertility". Arid Land Research and Management 26: 344–354.

59. Walker LR, Thompson DB, Landau FH (2001) Experimental manipulations of fertile islands and nurse plant effects in the Mojave Desert, USA. Western North American Naturalist 61: 25–35.

Chapter 5

A HYDROLOGICAL CONCEPT INCLUDING LATERAL WATER FLOW COMPATIBLE WITH THE BIOGEOCHEMICAL MODEL FOR SAFE

Giuliana Zanchi[1], Salim Belyazid [2], Cecilia Akselsson [1], Lin Yu [3], Kevin Bishop [4], Stephan J. Köhler [4], and Harald Grip [5]

[1]Department of Physical Geography and Ecosystem Science, Lund University, Sölvegatan 12, SE-223 62 Lund, Sweden

[2]Department of Physical Geography, Stockholm University, Svante Arrhenius väg 8, SE-114 18 Stockholm, Sweden

[3]Centre for Environmental and Climate Research, Lund University, Sölvegatan 37, SE-223 62 Lund, Sweden

[4]Department of Aquatic Sciences and Assessment, SLU, P.O. Box 7050, SE-750 07 Uppsala, Sweden

[5]Department of Forest Ecology and Management, SLU, Surbrunnsgatan 4, SE-114 27 Stockholm, Sweden

ABSTRACT

The study presents a hydrology concept developed to include lateral water flow in the biogeochemical model ForSAFE. The hydrology concept was evaluated against data collected at Svartberget in the Vindeln Research Forest in Northern Sweden. The results show that the new concept allows simulation of a saturated and an unsaturated zone in the soil as well as water flow that reaches the stream comparable to measurements. The most relevant differences compared to streamflow measurements are that the model simulates a higher base flow in winter and lower flow peaks after snowmelt. These differences are mainly caused by the assumptions made to regulate the percolation at the bottom of the simulated soil columns. The capability for simulating lateral flows and a saturated zone in ForSAFE can greatly improve the simulation of chemical exchange in the soil and export of elements from the soil to watercourses. Such a model can help improve the understanding of how environmental changes in the forest landscape will influence chemical loads to surface waters.

INTRODUCTION

Forests regulate the availability and quality of water resources. They generally reduce the total runoff through canopy interception and transpiration. Forests also change the velocity of the flow by enriching the soil with organic matter and changing the structure of the soil with their root system. By reducing the runoff and altering nutrient cycles, forests also change the quality of the soil water and runoff [1,2,3]. Therefore, environmental changes affecting forest dynamics, such as climate change or management changes, also affect water resources in forest catchments and eventually the loads of chemicals in inland and coastal waters [4, 5,6,7,8].

Biogeochemical models can be used to evaluate the effects of environmental changes on forest ecosystems [8,9,10,11,12]. They can include the simulation of hydrology of the system, but the level of complexity of the water flows varies significantly among models [13,14]. To succeed in simulating how nutrients are transported to rivers and surface waters, biogeochemical models need to include vertical and lateral flows in the soil.

The biogeochemical model ForSAFE is a mechanistic model designed to simulate the dynamic responses of forest ecosystems to environmental changes [11,15,16,17]. The hydrology module in ForSAFE was designed to simulate processes at the forest site level. It included the vertical flow of water in the soil profile but not lateral water flows.

The aim of this paper is to present the hydrology module developed to include lateral water flow, with the future aim of including it in the ForSAFE model. The inclusion of lateral flow will allow the simulation of water and chemistry transport from the forest ecosystem to the stream. In addition, it will simulate a saturated zone which will better represent moisture content in the soil and the processes in deeper soil layers as well as in the riparian zone. Moreover, in the current For SAFE model, the simulation of dynamics linked to fast hydrological response is limited by the fact that processes are simulated on a monthly time step. For this reason, the new hydrology module simulates dynamics at the daily time step. The hydrology module was calibrated and evaluated against data collected at Svartberget in the Vindeln Research Forest in Northern Sweden. The scope of the test in Svartberget was a proof of the concept rather than providing a best-fit to the measurements of water storage and flows by optimizing model parameters.

MODEL DESCRIPTION

The improved hydrology concept including lateral flow was designed to be compatible with the current structure of the For SAFE model.

For SAFE is a mechanistic model of the dynamics of forest ecosystems that combines chemical, physical and physiological processes [11, 15]. It integrates the three basic material and energy cycles in a single model: the biological cycle representing the processes involved in tree growth; the biochemical cycle including uptake, litter decomposition and soil nutrient dynamics; and the geochemical cycle including atmospheric deposition and weathering processes.

In For SAFE, each modelled forest unit includes a single soil column with different layers denoting the soil horizons. At present, the soil moisture content is calculated on a monthly basis for each of the defined soil layers. The hydrology module only simulates the vertical flows of water, including evapotranspiration and percolation. To include lateral water flow, it was necessary to represent the differences of soil properties not only with depth, but also along the slope. This was achieved by arranging several soil columns next to each other to discretely represent the vertical and horizontal structure of the soil in a forest transect. In order to test and better illustrate the new concept, the hydrology module was developed in the visual modeling environment STELLA (Figure 1) [18].

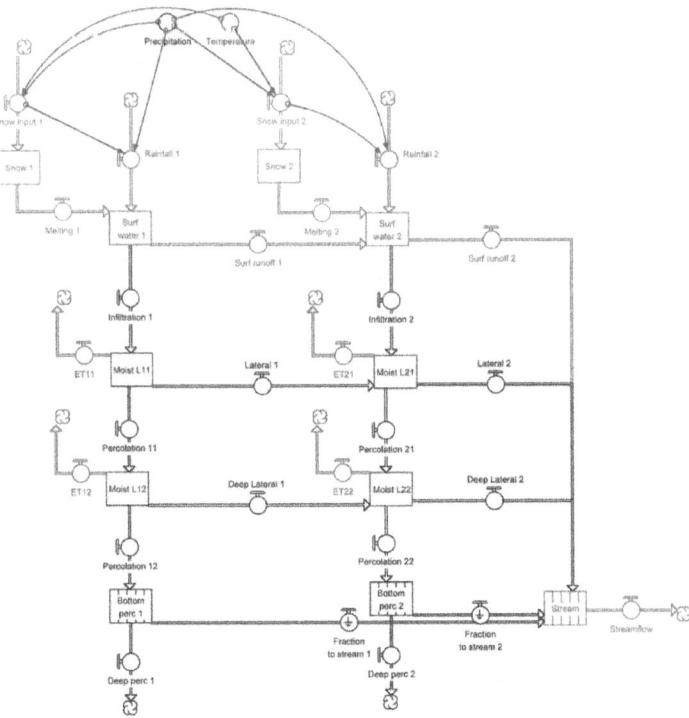

Figure 1: Simplified illustration (considering only two soil columns) of the hydrology

as conceptualized in STELLA. The boxes represent water stocks in elements storing water, the blue arrows represent water flows from and to water stocks, the stand-alone circles are converters containing data or equations and the red arrows are connectors indicating which factors are regulating the flows. The striped boxes represent conveyors that transfer water after a specified number of time units.

The main changes compared to the original module in For SAFE are:

- Increased time resolution from monthly to daily time steps

- The water flow from a given soil layer is constrained by the amount of water that receiving layers can accept vertically and horizontally

- The water flow is given a velocity regulated by the soil conductivity, controlling the amount of water that can move within each time step (per day)

- Inclusion of water movement along a slope, *i.e.*, surface runoff and lateral flow

- The soil hydraulic properties are assessed as a function of soil texture

The following sections describe in detail how the water flows and stocks are calculated in the new hydrology module as compared to that in For SAFE.

Water Inputs to the Soil

Water enters the system as precipitation on a daily time step (Figure 2, Water inputs). When entering the soil, the precipitation (Prec, $m \cdot d^{-1}$ of water) goes first through the same snow routine as in the For SAFE model. The precipitation is stored entirely in a snowpack (Snowin, $m^3 \cdot d^{-1}$ of water equivalents) when the average daily air temperature (Tm, °C) is below −5 °C. Between −5 °C and 2 °C, it is assumed that water inputs are a mix of snowfall and rainfall and only the snowfall is stored in the snowpack. The fraction of snowfall is reduced linearly with increasing air temperature until the 2 °C point, when all the precipitation is assumed to be in the form of rainfall [19].

$$\text{Snow}_{in} = \text{MAX}\left(0, \text{MIN}(\frac{T_m - 2}{-7}, 1)\right) \times \text{Prec} \times A_{1,j}$$

(1)

Where $A_{1,j}$ is the area of the first layer in soil column j. Generally, in the model, the first index of a variable (i) represents the number of the layer, starting at the soil surface and the second index (j) the number of the column, starting from the column next to the stream.

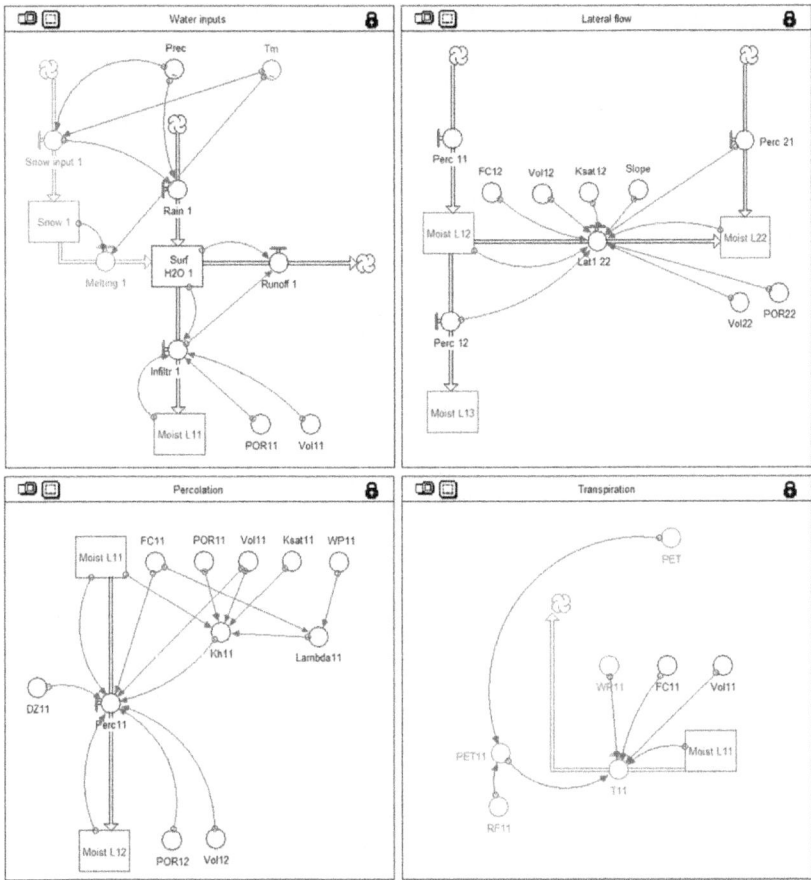

Figure 2: Illustration in STELLA of the main processes regulating water flows in the model: water inputs to the soil, percolation, lateral water flow and transpiration from a soil layer. The connectors (red arrows) indicate which factors are included in the calculation of the flows.

When the average daily temperature is greater than 0 °C, the water stored in the snowpack starts melting and enters the soil. The melting rate is determined by a degree-day factor (CMELT, m·°C^{-1}·d^{-1}) [19]. The sum of rainfall and snow melt at each time step (Surf H2O, m^3 water) corresponds to the maximum amount of water that can infiltrate in the upper soil layer. The actual infiltration in the first layer of each soil column j (Infilt$_{1,j}$, m^3·d^{-1} water) is limited by the level of saturation of the soil and cannot be greater than the volume of empty soil pores in the upper layer.

$$\text{Infilt}_{1,j} = \text{MIN}\left(\text{SurfH2O}, \left(\text{POR}_{1,j} - \text{Moist}_{1,j}\right) \times \text{Vol}_{1,j}\right)$$

(2)

Where $POR_{i,j}$ (m³ water m⁻³ soil) is the volume of pores per unit of soil volume, $Moist_{i,j}$ (m³ water m⁻³ soil) is the volume of water per unit of soil volume and $Vol_{i,j}$ (m³ soil) is the volume of the upper layer. The total volume of a soil layer is given by the thickness ($\Delta z_{i,j}$, m), the length ($\Delta x_{i,j}$, m) and the width of the layer ($\Delta y_{i,j}$, m). The impediment created by the coarse fragments is considered by reducing the total volume proportionally to the coarse fraction ($CF_{i,j}$):

$$Vol_{i,j} = \Delta z_{i,j} \times \Delta x_{i,j} \times \Delta y_{i,j} \times (1 - CF_{i,j}) \tag{3}$$

As a consequence, the presence of coarse fragments such as boulders or stones reduces the amount of pores in the soil and the water volume that can be retained in each soil layer.

The water that cannot infiltrate because the soil is saturated becomes surface runoff, and is directed towards the next column downhill or, in the case of the last column downstream, to the stream.

Water outputs from the soil

The water can leave each soil layer through three processes:

1. Percolation

2. Lateral flow

3. Transpiration

The percolation and lateral flows are, respectively, the vertical and horizontal water movements between soil layers, while transpiration is the water extracted by plants through their roots. The flows in the model are not regulated by a gradient of potential as determined by the laws of physical hydrology. The amount of water flowing in the soil at each time step is dependent on the hydraulic properties of the soil layers, as explained in the following sections.

The hydraulic properties of the soil used to assess the water outputs are:

1. the porosity, the fraction of pores in the soil volume (POR, m³ water m⁻³ soil)

2. the field capacity, the water content held in the soil when free drainage by gravity has stopped (FC, m³ water m⁻³ soil)

3. the permanent wilting point, the soil moisture content at which plants cannot extract more water from the soil (WP, m³ water m⁻³ soil)

4. the saturated hydraulic conductivity, the rate at which water moves through the pores in the saturated soil (Ksat, m·d⁻¹)

5. the unsaturated hydraulic conductivity, the rate at which water moves through the pores in the unsaturated soil (Kh, $m \cdot d^{-1}$)

6. the slope of the soil moisture characteristic, which is the relation between water tension and volumetric water content (λ)

The soil hydraulic properties are calculated for each soil horizon from the texture and the density of the soil, based on pedo-transfer functions [20, 21, 22].

All the hydraulic properties are representative for the volume occupied by the fine earth, *i.e.*, soil particles smaller than 2 mm in diameter and pores.

Percolation

The downward vertical flow is known as percolation and it concerns the volume of soil water above field capacity, *i.e.*, the water that can be moved by the force of gravity.

As in For SAFE, the maximum volume of water that can percolate from a soil layer ($Perc_{i,j}$, $m^3 \cdot d^{-1}$) is the water content above field capacity [19] (Figure 2, Percolation). In the new hydrology module, the percolation from layer i to layer $i + 1$ in soil column j is given by:

$$Perc_{i,j} = \begin{cases} MIN\left((Moist_{i,j}-FC_{i,j}) \times MIN\left(\frac{Vol_{i,j}}{dt}, Kh_{i,j} \times Av_{i,j}\right), EP_{i+1,j}\right), & Moist_{i,j} > FC_{i,j} \\ 0, & Moist_{i,j} \leqslant FC_{i,j} \end{cases}$$

(4)

Where $Moist_{i,j}$ (m^3 water m^{-3} soil) is the soil moisture content, $FC_{i,j}$ (m^3 water m^{-3} soil) the field capacity, $Vol_{i,j}$ (m^3) the volume of the percolating layer, $Kh_{i,j}$ ($m \cdot d^{-1}$) the unsaturated hydraulic conductivity and $Av_{i,j}$ (m^2) the cross-sectional area in the direction of the vertical flow. $EP_{i+1,j}$ is the capacity of the underlying layer given by the volume of empty pores at each time step (dt).

$$EP_{i+1,j} = MAX\left(\frac{(POR_{i+1,j}-Moist_{i+1,j}) \times Vol_{i+1,j}}{dt}, 0\right)$$

(5)

The unsaturated hydraulic conductivity (Kh, $m \cdot d^{-1}$) is calculated according to the function by Saxton and Rawls [21].

$$Kh_{i,j} = \left(Ksat_{i,j} \times \left(\frac{Moist_{i,j}}{POR_{i,j}}\right)^{3+\frac{2}{\lambda}}\right)$$

(6)

Where λ is the inverse of the slope of logarithmic tension-moisture curve:

$$\lambda = \frac{\ln(FC_{i,j}) - \ln(WP_{i,j})}{\ln(1500) - \ln(33)}$$

(7)

The percolation at the bottom of the modelled soil column can be divided into a fraction reaching the stream and a fraction that does not contribute to the streamflow (Figure 1). In addition, a transit time can be assigned to the bottom water flow, which can be used to better represent the time required to reach the stream.

Lateral Water Flow

In this model, the water flows both vertically and laterally downhill to the stream. When water above field capacity cannot percolate downwards because the layer below is saturated, it moves laterally to the next soil column downslope (Figure 2, Lateral Flow). That is, in the saturated zone and in the layers above the saturated zone, the flow is mainly horizontal. The vertical flow can also be constrained by a low permeability at the bottom of the soil column. This is simulated by reducing the conductivity that regulates the percolation from the deepest layer. The way flows are regulated in the model does not account for conditions when the flow is directed first laterally than vertically. Following Darcy's law, this would be the case when differences of potential are greater laterally than vertically, such as when the soil downhill is much drier or has a much finer texture than the deeper soil layers.

The lateral flow ($Lat_{i,j}$, $m^3 \cdot d^{-1}$) from a layer i in column j to a layer i in column $j-1$ (downstream) is given by:

$$Lat_{i,j} = \begin{cases} MIN\left((Moist_{i,j} - FC_{i,j}) \times MIN\left(\frac{Vol_{i,j}}{dt}, Ksat_{i,j} \times Ah_{i,j} \times Slp \right), EP_{i,j-1} \right), & Moist_{i,j} > FC_{i,j} \\ 0, & Moist_{i,j} \leq FC_{i,j} \end{cases}$$

(8)

Where $Ksat_{i,j}$ ($m \cdot d^{-1}$) is the saturated hydraulic conductivity, $Ah_{i,j}$ (m^2) the cross-sectional area in the direction of the horizontal flow and Slp the slope ($m \cdot m^{-1}$). $EP_{i,j-1}$ is the capacity of the receiving layer given by the volume of empty pores in the next column downslope at each time step:

$$EP_{i,j-1} = MAX(0, (POR_{i,j-1} - Moist_{i,j-1}) \times Vol_{i,j-1})$$

(9)

Transpiration

The assessment of the transpiration ($T_{i,j}$, $m^3 \cdot d^{-1}$ of water) from a soil layer i in soil column j is given by (adapted from [19]):

$$T_{i,j} = PET \times RF_{i,j} \times \left(MIN\left(MAX\left(0, \frac{Moist_{i,j} - WP_{i,j}}{LP_{i,j} - WP_{i,j}} \right), 1 \right) \right)$$

(10)

Where PET is the potential evapotranspiration, $RF_{i,j}$ the root fraction, $LP_{i,j}$ is the limit for potential evapotranspiration and $WP_{i,j}$ the permanent wilting point of the soil layer (Figure 2, Transpiration).

The PET in For SAFE is driven by tree photosynthesis, which in turn is dependent on nutrient availability. Since the tree growth processes and nutrient cycling are not modelled in this study, we adopted a simplified approach to estimate PET by calculating it as a function of temperature with the HBV model [23]. As in For SAFE, the potential plant transpiration is constrained by water availability in the soil. In addition, the transpiration equals the PET above a critical point identified by the limit of evapotranspiration, $LP_{i,j}$ which ranges between 50%–80% of the field capacity of the layer [24] The water availability is calculated as a function of the relative water content in the root-zone [19,24]. Moreover, the water that can be extracted from each layer is dependent on the amount of roots in that layer. In this study, the root fraction is assessed according to the data reported by Rosengren and Stjernquist [25] for different tree species. It is also assumed that the tree roots extend to soil layers up to 1 m depth.

Capillary rise due to surface tension is not considered in the model and this could limit the water available for trees when dry conditions persist for longer periods.

MODEL TEST

Site Description

The STELLA model was tested on the S-transect in the Vindeln Research Forest (64°14'N 19°46'E) in northern Sweden [26] by simulating water flows and soil moisture on a daily basis.

The S-transect is located at 250 m a.s.l. in the Krycklan catchment which covers a surface of about 68 km^2 (Figure 3). The transect is aligned parallel to lateral flow paths towards the Västrabäcken stream. Measurements of soil moisture, groundwater level and soil water chemistry have been collected from 1997 [26, 27]. The mean air temperature for the period 1981–2012 was 1.8°C, the average annual precipitation 631 mm and the average runoff 308 mm [26]. About half of the runoff occurs during a few weeks after the snowmelt. The vegetation is dominated by a mature stand of Norway spruce close to the stream and Scots pine upslope. The bedrock is gneiss and it is overlaid by glacial till, which is about 10–20 m deep in the lower part of the Svartberget catchment and 30–40 m in the middle part of the catchment. Nearest to the bedrock there is dense basal till that is overlaid by a less dense melt-out till of about 1–3 m depth [28]. The soil in the S-transect is an organic rich Histosol within 10 m of the Västrabäcken stream, with well-developed Podzols further upstream.

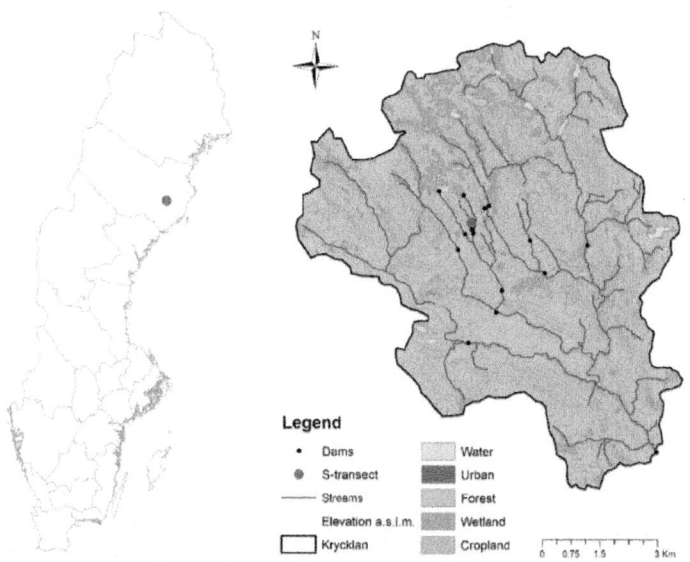

Figure 3: Location of the S-transect in Sweden and in the Krycklan catchment.

Transect Model

The soil transect was modelled as series of soil columns (C) representing the vertical and horizontal structure of the soil from the stream to the water divide.

The transect was designed as seven soil columns of different length (C1 to C7) and 1.5 m depth to represent discretely the change of soil type from the riparian zone to the mineral soil upstream (Figure 4). Data on grain size distribution at different points along the transect were used to define the texture of the seven soil columns, each divided in seven layers ([29], Table S1).

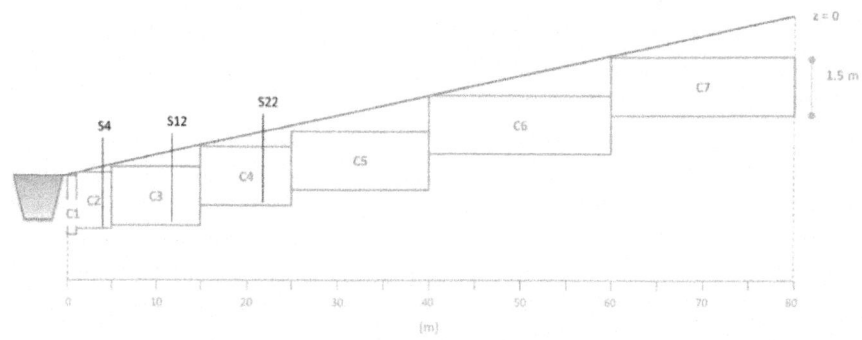

Figure 4: Representation of the S-transect in the hydrology model when the water

divide is assumed at 80 m from the stream. C: modelled soil column. S: point of measurement. The number of the sampling points represents the distance from the stream (4, 12 and 22 m).

Measurements of daily precipitation, average temperature and grain size distribution were used as inputs to the model. Model parameters and results were tested against hydrological measurements taken at three different stations along the transect which are located at 4, 12 and 22 m from the stream (S4, S12, S22) (Figure 4).

Parameterization of Soil Hydraulic Properties

Measurements of soil hydraulic properties are not commonly available. Therefore, the model is structured to calculate them based on soil texture—an input required also by the ForSAFE model- and the pedo-transfer functions [20]. In this study, the soil hydraulic properties were assessed for the different layers of all soil columns based on texture data collected along the entire transect (Table S2).

The functions were tested against measurements available along the transect. Measurements of hydraulic properties were available only at the three sampling point S4, S12 and S22 at selected depths. Porosity and the soil moisture characteristic (pF curves) were measured on cylinder samples of the soil using the pressure plate method, except for the 150 m tension for which a pressure chamber was used [27]. The saturated hydraulic conductivity was measured with a double-ring infiltrometer [30]. Texture data from the same or comparable sampling points were used to estimate the hydraulic properties with the pedo-transfer functions.

The calculated porosity is generally in good agreement with the measured data (Figure 5, POR). Some discrepancies are observed in organic soil layers where the porosity is high due to the high organic matter content [31, 32, 33]. A possible reason for this difference is that the volumetric water content can vary substantially within short distances at the interface between organic and mineral soil, due to the large difference in physical properties between those two types of soil. When the estimated porosity in a layer is higher than the actual, water content at saturation will be overestimated by the model. This could have the effect of overestimating the water stored in the soil, thus underestimating the streamflow. The opposite will happen when the calculated porosity is underestimated.

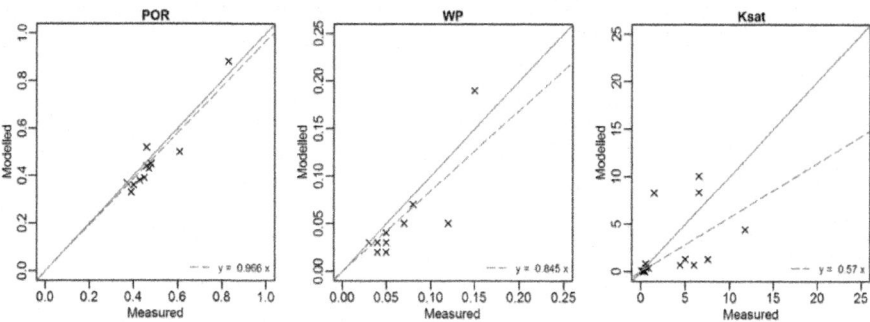

Figure 5: Testing of estimated hydraulic properties against measurements at the sampling points S4, S12, S22 of the S-transect. POR: porosity; WP: permanent wilting point; Ksat: saturated hydraulic conductivity. The dotted lines and their slope are an indication of the discrepancy between estimates and measured data. When the slope is > 1 the measured data are higher than the estimates and when <1 measurements are lower than estimates.

The estimated permanent wilting point is also comparable to measurements (Figure 5, WP). When different, the water volume retained at the wilting point is a low fraction of the soil volume. Therefore, the difference in terms of water content is small and it should have a minor impact on the estimate of water storage and fluxes.

Greater discrepancies are found between the calculated saturated hydraulic conductivity and measurements (Figure 5, Ksat). In upper layers, the calculated Ksat is generally higher than the empirical data and lower in deeper layers. Therefore, in agreement with other studies [34, 35], the results indicate that the estimated Ksat decreases more rapidly with depth and is higher in upper layers than measurements. These discrepancies will have an effect on the velocity of the flow: when the estimated Ksat is lower than the actual, water will reach the stream later than in reality. Both calculated values and measurements indicate that layers with high organic matter (>10%) have a lower saturated hydraulic conductivity.

Sensitivity Analysis

A sensitivity analysis was performed to evaluate a set of model parameters that could influence the model outputs. These were also the parameters that were considered for the model calibration. A one-factor-at-the-time method (OFAT) was used to test the model sensitivity to parameter perturbation [36]. Two objective functions were used in the analysis: Bias and Root Mean Square Error (RMSE) (Table S3). The functions evaluate the discrepancy between model results when parameters are varied within a certain range relative to a baseline value (Table 1). The ranges were defined based on available information from the site or literature, as explained in the following paragraphs.

Table 1: Parameter ranges and baseline values used in the sensitivity analysis.

Parameter	Description	Unit	Baseline	Range of Variation
WD	Transect length given by the distance from the river to the water divide	m	110	80–140
Kbott	Saturated hydraulic conductivity at the bottom of the soil column j, *i.e.*, in layer (j,7)	$m \cdot d^{-1}$	Ksat(j,7) × 0.5	0–Ksat(j,7)
FrStream	Fraction of bottom percolation reaching the stream	fraction	Linearly decreasing	0.8–1.0
Cmelt	Degree-day factor for snow melt	$mm \cdot {}^{\circ}C^{-1} \cdot d^{-1}$	1.5	1–2
LP	Limit for potential evapotranspiration, expressed as a fraction of FC	fraction	0.65 × FC(j,i)	0.5–0.8 × FC(j,i)
POR	Porosity of all soil layers	fraction	POR(j,i)	POR(j,i) ± 10%
Ksat	Saturated conductivity of all soil layers	$m \cdot d^{-1}$	Ksat(j,i)	Ksat(j,i) ± 10%

A Lidar-based digital elevation model was used to define the hillslope up to the water divide (WD). With the 1 m DEM, the transect extended 80 m to the water divide, while using a 5 m DEM realigned the transect slightly and pushing the water divide to 140 m [37]. The two different estimates of the water divide compared to a mean value of 110 m were used to test the effect of changes of hillslope length on the model results.

The model simulates the hillslope hydrology up to 1.5 m depth. Without further assumptions the percolation at the bottom of the column is only regulated by the calculated Ksat of the bottom layer (Kbott). As a consequence, the lower permeability in soil deeper than 1.5 m would not be considered in the simulation. The effect of changing the permeability of the bottom layer was tested in the sensitivity analysis by varying the Kbott within a range between zero (e.g., impermeable rock) and a value equal to the Ksat of the bottom layer (no further impediment at the bottom of the column).

A further unknown factor is the type of connection existing between the water percolating from the bottom of the columns and the river (Fr Stream). When evaluating the model, the simulated streamflow is compared to a measured value taken at a certain point along the river. Ideally all the water from the simulated area discharges at the point of measurement (Fr Stream = 1), but part of the deep flow can also discharge in the area downstream from the point of measurement [38]. Since the transect is aligned parallel to lateral flow paths, most of the bottom percolation should contribute to the stream flow measured at the dam downstream. In the sensitivity analysis the fraction to the stream was varied between 0.8 and 1. As a baseline scenario, we considered a linear decrease of the fraction to the stream from 1 in the column next to the river to 0.8 at the water divide. The change of the degree-day factor was used to test the sensitivity of the model to the simulation of snow melt (Cmelt). In

For SAFE a value of 1.5 mm °C^{-1}·d^{-1} was used in several studies in Sweden [39]. A variation of ±0.5 mm °C^{-1}·d^{-1} was included in the sensitivity analysis.

The effect of transpiration was evaluated by varying the limit for potential evapotranspiration, LP, that is calculated as a fraction of the FC, ranging from 0.5 to 0.8 of FC [24, 40].

Other factors that can affect model results are the estimated hydraulic properties, which determine the amount of water stored and flowing in the soil. In the sensitivity analysis we tested the effects of varying the porosity (POR) and the saturated conductivity (Ksat) of ±10% of the calculated value. According to the pedo-transfer functions the change of porosity affects also the field capacity and the wilting point (on average ±12% and ±14%, respectively).

Model results are also affected by the uncertainty of the input data, such as texture, PET and climate data, but these effects were not included in the sensitivity analysis.

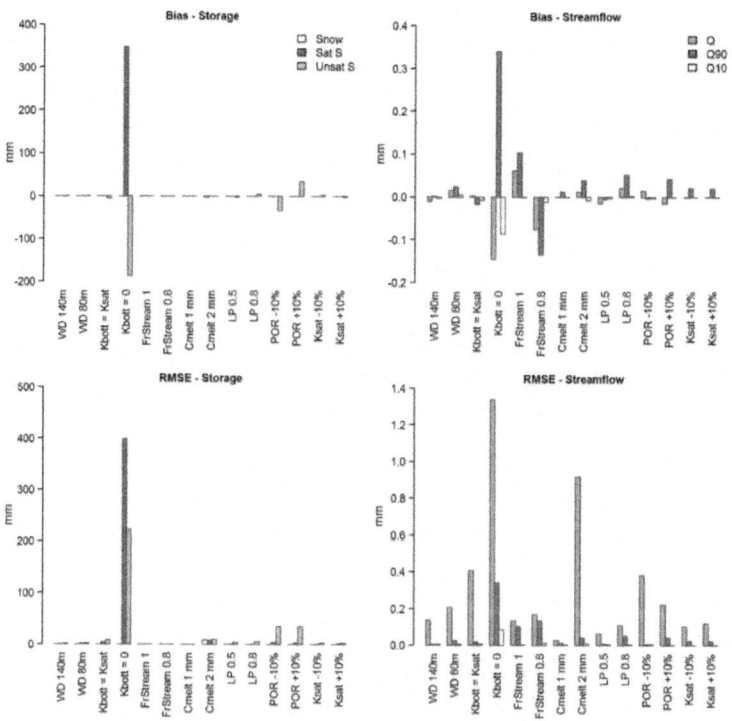

Figure 6: Change of water storage and streamflow components when parameters are varied compared to a baseline. The change is expressed in terms of Bias and RMSE.

We evaluated the model sensitivity by analyzing the modelled water storage and streamflow (Table 2 and Figure 6). We considered three different components of the water storage: the snowpack, the saturated and the unsaturated storages in the soil. We also analyzed the changes in daily streamflow (Q) and the high and low peak values by comparing the 90% and 10% percentile values of daily streamflow.

Table 2: Results of the sensitivity analysis. The values represent the difference in mm of water compared to the baseline scenario.

Scenario	Range	Snow		Saturated Storage		Unsaturated Storage		Streamflow, Q		Q 90% perc (Q90)		Q 10% perc (Q10)	
		Bias	RMSE	Bias	RMSE	Bias	RMSE	Bias	RMSE	Bias	RMSE	Bias	RMSE
WD	140 m	0.00	0.00	0.00	0.40	−0.53	1.08	−0.01	0.14	0.00	0.00	0.00	0.00
	80 m	0.00	0.00	0.01	0.76	0.78	1.71	0.02	0.21	0.02	0.02	0.01	0.01
Kbott	Ksat(j,7)	0.00	0.00	−0.31	4.05	−6.02	7.59	0.00	0.41	−0.02	0.02	−0.01	0.01
	0	0.00	0.00	348.27	399.01	−186.35	223.42	−0.15	1.34	0.34	0.34	−0.09	0.09
FrStream	1	0.00	0.00	0.00	0.00	0.00	0.00	0.06	0.14	0.10	0.10	0.00	0.00
	0.8	0.00	0.00	0.00	0.00	0.00	0.00	−0.08	0.17	−0.14	0.14	−0.01	0.01
Cmelt	1 mm	0.00	0.00	0.00	0.00	0.00	0.00	0.00	0.03	0.01	0.01	0.00	0.00
	2 mm	−2.89	7.39	−0.07	6.97	−0.22	9.32	0.01	0.92	0.04	0.04	−0.01	0.01
LP	0.5 × FC	0.00	0.00	0.00	0.15	−2.48	3.79	−0.02	0.06	−0.01	0.01	0.00	0.00
	0.8 × FC	0.00	0.00	0.00	0.63	3.56	5.28	0.02	0.11	0.05	0.05	0.00	0.00
POR	−10%	0.00	0.00	0.17	3.19	−34.49	34.83	0.01	0.38	0.00	0.00	0.00	0.00
	10%	0.00	0.00	−0.10	2.29	34.49	34.71	−0.02	0.22	0.04	0.04	0.00	0.00
Ksat	−10%	0.00	0.00	0.03	0.54	2.37	2.63	0.00	0.10	0.02	0.02	0.00	0.00
	10%	0.00	0.00	0.00	1.29	−2.04	2.55	0.00	0.12	0.02	0.02	0.00	0.00

A reduction of the conductivity regulating the bottom percolation has the most significant impact on both the simulated water storage and the streamflow. By reducing the Kbott, it is possible to increase the saturated storage in the soil, *i.e.*, simulate a higher groundwater level. In addition, a reduced permeability at the bottom significantly changes the streamflow, which is reduced on average on a daily basis, but increased at high flow.

To a minor extent, the water storage is also affected by the calculated porosity: with higher porosity the soil is on average more unsaturated.

Other relevant factors for the simulation of streamflow are:

- The fraction of bottom percolation to the stream (frstream): when increased the overall streamflow increases, as indicated by the Bias;

- An increase of snow melting factor (Cmelt) affects significantly the distribution of the flow during the year (increase of RMSE), but has a minor effect on the overall water balance.

Calibration

As a following step, the model results obtained when applying baseline parameter values were compared against measured data describing soil moisture and streamflow. These included dynamic soil moisture measurements with TDR and the streamflow data at the Västrabäcken stream water site (site 2, [26]) from the period 2011–2014. Soil TDR were available at 5, 12 and 22 m from the stream which are comparable to columns C2, C3 and C4 in the model. The TDR data of the second and fourth layer were available for all the three measurement points and therefore compared to the modelled results (Table 3). The discrepancies between modelled results and measurements were compared in terms of Bias, RMSE and normalized mean error (NME) (Table S3). Almost all the parameters are significantly underestimated. By adopting baseline parameter values, the soil is most of time unsaturated and the streamflow on a daily basis and at high peak flows lower than measurements. The only parameter that is greatly overestimated is the 10% percentile of the streamflow, *i.e.*, the low flow values.

Table 3: Comparison of modelled results to measurements when adopting baseline parameter values. C: soil column; L: layer number. For Bias and RMSE, soil moisture data are in m³ water m⁻³ soil and streamflow data in mm.

		Bias	RMSE	NME (%)
Soil moisture	C2:L2	−0.101	0.242	34.8
	C2:L4	−0.194	0.196	35.2
	C3:L2	−0.163	0.239	46.6
	C3:L4	−0.268	0.270	53.3
	C4:L2	0.007	0.083	26.9
	C4:L4	−0.139	0.163	44.0
Streamflow	Q	−0.022	1.387	108.6
	Q90	−0.664	0.664	37.4
	Q10	0.103	0.103	683.1

This comparison suggests that a high conductivity at the bottom of the soil columns compromises the simulation of a saturated zone and releases a constant base flow which is much higher than reality.

Therefore, in the calibration phase, the Ksat of the bottom layer was reduced by two orders of magnitude. A further decrease of Ksat was necessary in column C4 (15–25 m from the stream) where the calculated Ksat was at least twice as large as in the other columns. As a result, a total decrease of 2×10^{-2} was applied n C4 and the Ksat at the bottom of the modelled transect was reduced from 0.33 m·d⁻¹ to 0.003 m·d⁻¹ on average. The adjustment was necessary to avoid an excessive loss of water through bottom percolation that would compromise the simulation of deep saturated layers, *i.e.*, of the groundwater level. We justify this choice by assuming that a very low conductivity at the bottom of the column would include the lower permeability in soil deeper than 1.5 m that is not considered in the simulation.

The results from the calibration are evaluated in the following sections by analyzing the components of the water storage and of the water flows. The results on the dynamic soil moisture and runoff are compared against the TDR and streamflow measurements on a daily basis.

MODEL RESULTS AND DISCUSSION

Simulated Water Storage and Fluxes

The runoff (R) in the entire simulation period is 36% of the precipitation (P) (Table 4). A smaller share of the runoff (11%) is simulated as deep percolation and does not reach the stream, while most of it contributes to the stream flow. Transpiration plays a significant role in the water balance by using about 60% of the total precipitation. Over the entire simulation period, the water storage in the soil increases by 4%. The increase of storage mostly happens in the first year of simulation, because it is assumed that all layers have an initial water content equal to field capacity. Therefore, a period of stabilization of the model is required to reach the saturation level in deeper soil layers.

Table 4: Components (mm·y^{-1}) of the water balance at the study area. T: transpiration; R: total runoff (including deep percolation and stream flow); ΔS: change of soil water storage; P: precipitation.

Year	T	Deep Perc	Stream Flow	R	ΔS	T + R + ΔS	P
2011	368	3	52	55	225	648	649
2012	451	14	255	270	108	829	829
2013	441	15	250	265	−64	641	642
2014	412	12	203	215	−89	538	537
Total	1570	105	863	968	119	2657	2657
% of P	59%	4%	32%	36%	4%	100%	

Figure 7 illustrates the magnitude and the changes of the water storage over time. The storage is divided into storage in the saturated and unsaturated zones and in the snow pack. The period of stabilization required to reach the saturation level in deeper soil layers is clearly shown by the absence of a saturated zone at the beginning of the simulation. The stabilization ends some months after the first snowmelt, when the soil starts receiving water inputs. In the following years (2012–2014), the model simulates a seasonal change of the storage which reflects the accumulation of water in the snow pack, the increase of soil water storage after snow melt, both in the unsaturated and saturated zone, and the decrease of soil moisture in summer, due to plant transpiration.

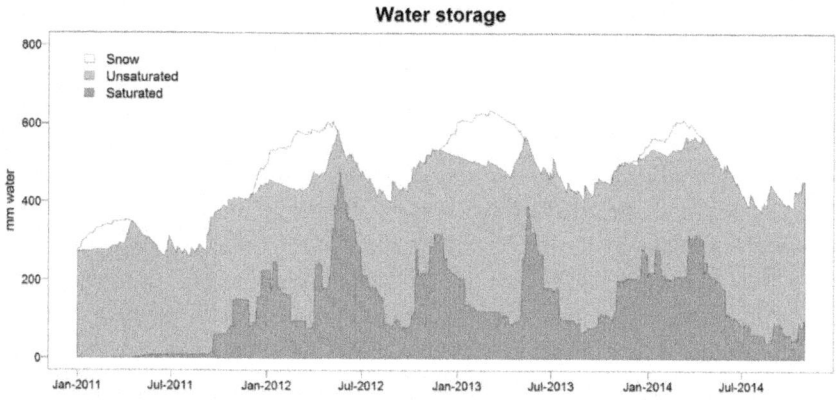

Figure 7: Water storage in the saturated and unsaturated zone in the soil and in the snow pack over the simulation period.

Soil Moisture

Figure 8 compares the modelled soil moisture to measured TDR data at different depths at the three sampling points S4, S12, S22. The figure shows the simulation of soil moisture in the period 2012–2014, *i.e.*, excluding the period of stabilization.

The model simulates permanent saturated layers at depths similar to measurements and levels of soil moisture which are comparable to the TDR data.

However, some of the simulated moisture dynamics differ significantly from the measurements. The TDR measurements in the upper layers show a decrease of moisture in winter that reaches values below field capacity (e.g., in layers up to 20 cm depth). In the same period, the model predicts water contents at the lowest close to field capacity, since transpiration does not occur in winter. The decrease of the TDR values is mainly caused by soil and water frost in the upper soil layers, which is not shown in modelled results, since frozen water is not simulated by the model.

Other discrepancies between model and measurements can be attributed to different soil hydraulic properties at the points of measurement and in the modelled soil layer. Due to its structure, the model assumes that all soil properties are homogeneous at any given point within a soil layer, as compared to measurements, which represent a specific point in depth and in distance from the stream. When modelled and measured saturation levels differ, the calculated porosity for the entire layer is most likely different from the porosity at the point of measurement.

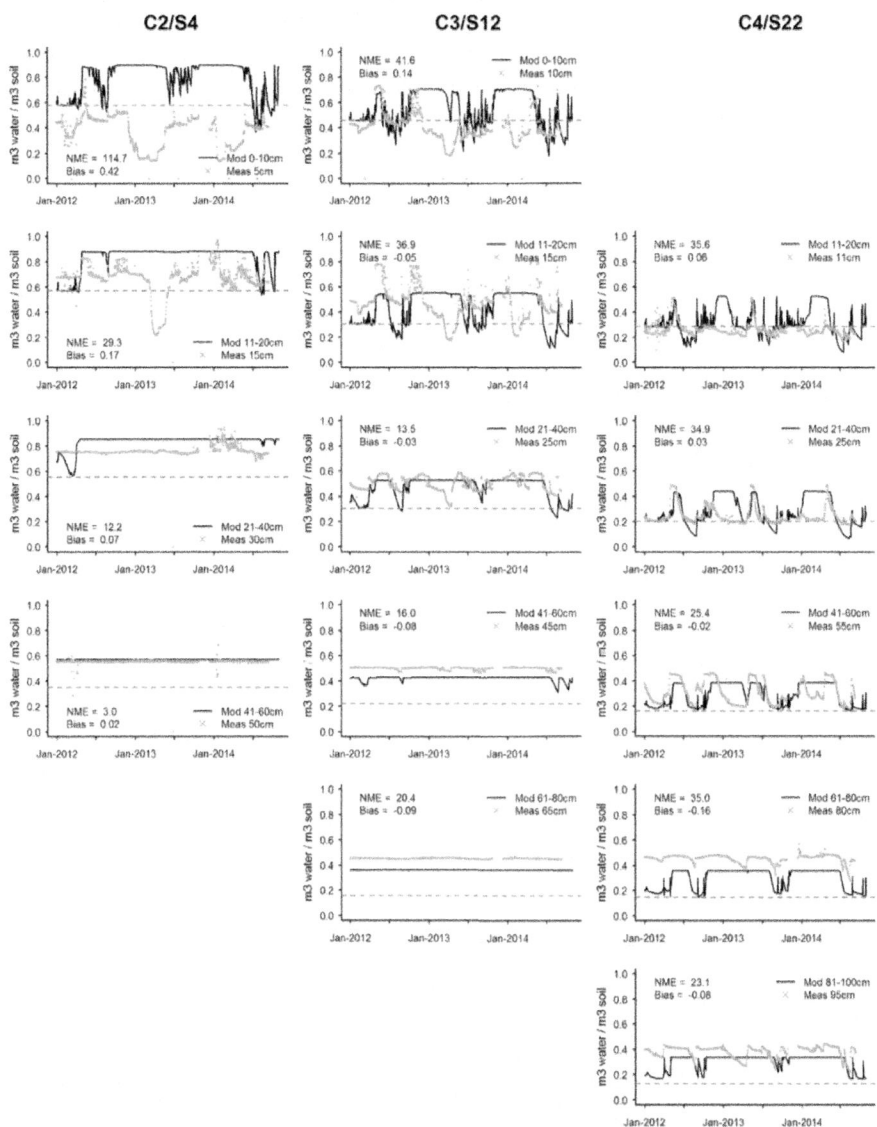

Figure 8: Modelled soil moisture as compared to the TDR data at increasing distance from the stream. The three modelled soil columns up to 25 m from the stream (C2, C3, C4) are compared to the measurements at the sampling points at 4, 12 and 22 m (S4, S12, S22). For each soil profile, simulations and measurements are compared at different soil depths. Modelled layers of similar depths are compared in the same row. Measurements were not available for layers that are missing. Differences between model and measurements are evaluated in terms of NME (%) and Bias (m³·m⁻³). Dashed grey lines represent the water content at field capacity.

Moreover, the parameterization of the soil hydraulic conductivity at the bottom of the soil columns can result in soil permeability different from the actual one in the deep soil. The difference can produce some of the discrepancies between the fluctuations of modelled soil moisture and of the TDR data in deeper layers.

Streamflow

The comparison of the modelled and measured streamflow shows that the model tends to overestimate the annual water flow at the stream (Figure 9). In 2013–2014, the cumulative modelled streamflow over time is about 10% higher than measurements.

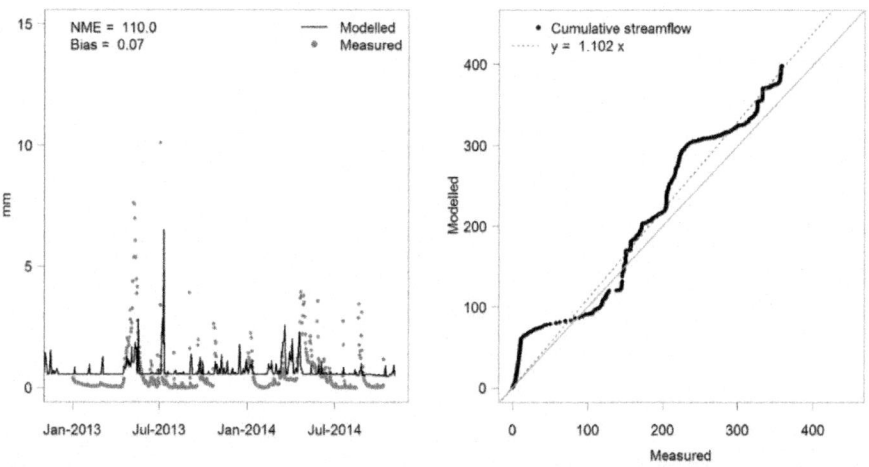

Figure 9: Modelled as compared to measured streamflow. Differences between model and measurements are evaluated in terms of NME (%) and Bias (mm). The graph on the right compares the cumulative streamflow in 2013-2014 when measurements were available.

The overestimation of the annual streamflow is mainly caused by the modelled base flow which is higher than the measured one: about 0.60 mm·d^{-1} against the observed 0.25 mm·d^{-1} in winter. As highlighted by the sensitivity analysis, the magnitude of the modelled base flow is influenced by the assumptions made to model the bottom percolation: a reduced Ksat at the bottom of the soil columns have the effect of reducing the overall streamflow and the 10% percentile value. In addition, the model does not simulate soil and water frost, as shown by the mismatch between measured and modelled soil moisture in winter. The unfrozen soil water could contribute to the winter flow and thereby overestimate the streamflow in winter.

The results presented in Figure 9 also highlight that the model tends to underestimate the peak flows in spring. Also in this case, the high flow values are mostly affected by the Ksat at the bottom of the columns: a reduction of Ksat increases the 90% percentile value of the streamflow. Other factors that could contribute to the seasonal differences between modelled and measured streamflow are a snow routine that does not fully capture the processes of snow accumulation and melting, the parameterization of plant transpiration and the lack of preferential flow paths in the model.

As shown in the sensitivity analysis, the parameterization of snow melting mainly changes the distribution of the flow and affects the peak flows: a faster snow melt anticipates and increases the peak flows by concentrating the release of water from the snow pack in a shorter time. Similarly, it can be expected that by modifying the parameters regulating snow accumulation, the yearly distribution of the streamflow would change. An increase of precipitation stored in the snowpack could reduce the amount of water discharged in winter and increase the streamflow in spring.

Transpiration affects the flow during the vegetation period. A lower transpiration could reduce the uptake in spring and contribute to simulate higher peak flows. However, the overall effect of the simulation of the streamflow would be less significant than the factors previously discussed.

The lack of preferential paths contributes to the underestimation of the conductivity of the soil at high precipitation events causing a water flow more distributed over time and thereby a possible underestimation of the discharge at high flow. The difference between modelled and real soil conductivity could have an effect on the residence time of water, *i.e.*, the time spent by water in the soil, which is relevant to predict the amount of chemicals transported to the stream. By calculating the mean residence time (T_R, days) as the ratio between the simulated soil water storage and the streamflow over the period 2012–2014, we found that the mean residence time is related to the streamflow (Sf, mm·d^{-1}) according to the following exponential function: $T_R = 496.72 \times Sf^{-0.899}$ ($R^2 = 0.93$); the median T_R is 846 days.

CONCLUSIONS

As compared to the previous hydrology module in ForSAFE, the new hydrology concept allows for the simulation of a saturated and an unsaturated zone in the soil, as well as a water flow that reaches the stream in a way that is consistent with measurements. It simulates with a good approximation the saturated zone and the level of soil moisture at different depths. In addition, it captures part of the flow peaks observed in the stream and it simulates a total annual streamflow comparable to the measured one.

The most relevant differences compared to measurements that will affect the chemical transport to the stream are that the model simulates a higher base flow in winter and lower flow peaks after snowmelt. The discrepancies are mainly caused by the assumptions made to regulate the percolation at the bottom of the soil columns. To simulate a saturated zone in the soil and a seasonal change of streamflow it is necessary to reduce the conductivity of the deepest soil layer to include to the lower permeability of soil below the simulated depth. The differences relative to measurements observed in this study will mainly have an effect on the timing with which water reaches the stream. The residence time of water in the soil affects the biogeochemical processes in the soil and eventually the amount of chemicals that are transported to the river.

The discrete structure of the model also limits its capability to represent water storage and transport which in reality are continuous. However, it makes the hydrology module suitable for coupling to the ForSAFE model. The simulation of lateral flows and a saturated zone in ForSAFE will couple several processes in the forest ecosystem with water flows to the steam. This will greatly improve the simulation of chemical exchange in the soil and chemical transport from the soil to watercourses.

Future work including chemical transport will help clarify the impact of water residence times and help identify possible improvements in the modeling of hydrology in ForSAFE.

Supplementary Material

Table S1: Description of the soil columns in the S-Transect model; Table S2: Calculated soil hydraulic properties for the modelled soil columns of the S-transect; Table S3: Objective functions.

Acknowledgments

The authors wish to thank for the financial support granted by the projects "ForWater" and "Managing Multiple Stressors in the Baltic Sea" funded by FORMAS. We also thank Nino Amvrosiadi for the valuable discussions that contributed to the model development and Martin Erlandsson for sharing the data on soil texture.

Author Contributions

The model was developed by Giuliana Zanchi and Salim Belyazid. Lin Yu and Harald Grip contributed to a further refinement of the model. Kevin Bishop and Stephan Köhler provided the experimental data used to test the model. All authors contributed to data analysis, paper writing and proofreading.

Conflicts of Interest

The authors declare no conflict of interest.

REFERENCES

1. Chang, M. *Forest Hydrology: An Introduction to Water and Forests*, 2nd ed.; CRC Press: Boca Raton, FL, USA, 2006.

2. Neary, D.G.; Ice, G.G.; Jackson, C.R. Linkages between forest soils and water quality and quantity. *For. Ecol. Manag.* **2009**, *258*, 2269–2281.

3. FAO. *Forest and Water-International Momentum and Action*; FAO: Rome, Italy, 2013; p. 86.

4. Muller, F.; Chang, K.-C.; Lee, C.-L.; Chapman, S. Effects of temperature, rainfall and conifer felling practices on the surface water chemistry of northern peatlands. *Biogeochemistry* **2015**, *126*, 343–362.

5. Hellsten, S.; Stadmark, J.; Pihl Karlsson, G.; Karlsson, P.E.; Akselsson, C. Increased concentrations of nitrate in forest soil water after windthrow in southern sweden. *For. Ecol. Manag.* **2015**, *356*, 234–242.

6. Taboada-Castro, M.M.; Rodríguez-Blanco, M.L.; Diéguez, A.; Palleiro, L.; Oropeza-Mota, J.L.; Taboada-Castro, M.T. Effects of changing land use from agriculture to forest on stream water quality in a small basin in nw spain. *Commun. Soil Sci. Plant Anal.* **2015**, *46*, 353–361.

7. Schelker, J.; Öhman, K.; Löfgren, S.; Laudon, H. Scaling of increased dissolved organic carbon inputs by forest clear-cutting—what arrives downstream? *J. Hydrol.* **2014**, *508*, 299–306.

8. Xiao, W.; Chun, Z.; Qing, J. Impacts of climate change on forest ecosystems in northeast china. *Adv. Clim. Chang. Res.* **2013**, *4*, 230–241.

9. Aber, J.; Neilson, R.P.; McNulty, S.; Lenihan, J.M.; Bachelet, D.; Drapek, R.J. Forest processes and global environmental change: Predicting the effects of individual and multiple stressors we review the effects of several rapidly changing environmental drivers on ecosystem function, discuss interactions among them, and summarize predicted changes in productivity, carbon storage, and water balance. *Bioscience* **2001**, *51*, 735–751.

10. Campbell, J.L.; Rustad, L.E.; Boyer, E.W.; Christopher, S.F.; Driscoll, C.T.; Fernandez, I.J.; Groffman, P.M.; Houle, D.; Kiekbusch, J.; Magill, A.H.; *et al.* Consequences of climate change for biogeochemical cycling in forests of northeastern north americathis article is one of a selection of papers from ne forests 2100: A synthesis of climate change impacts on forests of the northeastern us and eastern canada. *Can. J. For. Res.* **2009**,

39, 264–284.

11. Belyazid, S.; Westling, O.; Sverdrup, H. Modelling changes in forest soil chemistry at 16 Swedish coniferous forest sites following deposition reduction. *Environ. Pollut.* **2006**, *144*, 596–609.

12. Gaudio, N.; Belyazid, S.; Gendre, X.; Mansat, A.; Nicolas, M.; Rizzetto, S.; Sverdrup, H.; Probst, A. Combined effect of atmospheric nitrogen deposition and climate change on temperate forest soil biogeochemistry: A modeling approach. *Ecol. Model.* **2015**, *306*, 24–34.

13. Homann, P.S.; McKane, R.B.; Sollins, P. Belowground processes in forest-ecosystem biogeochemical simulation models. *For. Ecol. Manag.* **2000**, *138*, 3–18.

14. Maxwell, R.M.; Putti, M.; Meyerhoff, S.; Delfs, J.-O.; Ferguson, I.M.; Ivanov, V.; Kim, J.; Kolditz, O.; Kollet, S.J.; Kumar, M.; *et al.* Surface-subsurface model intercomparison: A first set of benchmark results to diagnose integrated hydrology and feedbacks. *Water Resour. Res.* **2014**.

15. Wallman, P.; Svensson, M.G.E.; Sverdrup, H.; Belyazid, S. Forsafe—An integrated process-oriented forest model for long-term sustainability assessments. *For. Ecol. Manag.* **2005**, *207*, 19–36. Yu, L.; Belyazid, S.; Akselsson, C.; van der Heijden, G.; Zanchi, G. Storm disturbances in a swedish forest—A case study comparing monitoring and modelling. *Ecol. Model.* **2016**, *320*, 102–113.

16. Zanchi, G.; Belyazid, S.; Akselsson, C.; Yu, L. Modelling the effects of management intensification on multiple forest services: A swedish case study. *Ecol. Model.* **2014**, *284*, 48–59.

17. Richmond, B. *An Introduction to System Thinking with Stella*; Isee Systems Inc: Lebanon, NH, USA, 2004.

18. Lindström, G.; Gardelin, M. Chapter 3.1: Hydrological Modelling-Model Structure. In *Modelling Groundwater Response to Acidification*; Swedish Meteorological and Hydrological Institute (SMHI): Norrköping, Sweden, 1992; pp. 33–36.

19. Balland, V.; Pollacco, J.A.P.; Arp, P.A. Modeling soil hydraulic properties for a wide range of soil conditions.*Ecol. Model.* **2008**, *219*, 300–316.

20. Saxton, K.E.; Rawls, W.J. Soil water characteristic estimates by texture and organic matter for hydrologic solutions. *Soil Sci. Soc. Am. J.* **2006**, *70*, 1569–1578.

21. Reid, I. The choice of assumed particle density in soil tests. *Area* **1973**, *5*, 10–12.

22. Bergström, S. *The HBV Model: Its Structure and Applications*; Swedish

Meteorological and Hydrological Institute (SMHI): Nörrköping, Sweden, 1992; p. 35.

23. Dingman, S.L. *Physical hydrology*, 2nd ed.; Waveland Press, Inc.: Long Grove, IL, USA, 2008; p. 646.

24. Rosengren, U.; Stjernquist, I. *Gå på djupet!: Om rotdjup och rotproduktion i olika skogstyper*; SUFOR: Alnarp, Sweden, 2004; p. 55.

25. Laudon, H.; Taberman, I.; Ågren, A.; Futter, M.; Ottosson-Löfvenius, M.; Bishop, K. The krycklan catchment study—a flagship infrastructure for hydrology, biogeochemistry, and climate research in the boreal landscape. *Water Resour. Res.* **2013**, *49*, 7154–7158.

26. Nyberg, L.; Stähli, M.; Mellander, P.-E.; Bishop, K.H. Soil frost effects on soil water and runoff dynamics along a boreal forest transect: 1. Field investigations. *Hydrol. Process.* **2001**, *15*, 909–926.

27. Lindqvist, G.; Nilsson, L.; Gonzalez, G. *Depth of Till Overburden and Bedrock Fractures on the Svartberget Catchment as Determined by Different Geophisical Methods*; University of Luleå: Luleå, Sweden, 1989; p. 16.

28. Erlandsson, M.; Uppsala University, Uppsala, Sweden. Personal communication, 2015.

29. Bishop, K.H. *Episodic Increases in Stream Acidity, Catchment Flow Pathways and Hydrograph Separation*; University of Cambridge: Cambridge, UK, 1991.

30. Yi, S.; Manies, K.; Harden, J.; McGuire, A.D. Characteristics of organic soil in black spruce forests: Implications for the application of land surface and ecosystem models in cold regions. *Geophys. Res. Lett.* **2009**.

31. Mitsch, W.J.; Gosselink, J.G. *Wetlands*, 5th ed.; Wiley: Hoboken, NJ, USA, 2015.

32. Kellner, E. *Wetland—Different Types, Their Properties and Functions*; Department of Earth Sciences/Hydrology: Uppsala, Sweden, 2003; p. 62.

33. Bishop, K.; Seibert, J.; Nyberg, L.; Rodhe, A. Water storage in a till catchment. Ii: Implications of transmissivity feedback for flow paths and turnover times. *Hydrol. Process.* **2011**, *25*, 3950–3959.

34. Rodhe, A. On the generation of stream runoff in till soils. *Nord. Hydrol.* **1988**, *20*, 1–8.

35. Cuo, L.; Giambelluca, T.W.; Ziegler, A.D. Lumped parameter sensitivity analysis of a distributed hydrological model within tropical and temperate catchments. *Hydrol. Process.* **2011**, *25*, 2405–2421.

36. Amvrosiadi, N.; Bishop, K.; Grabs, T.; Beven, K.; Seibert, J. Water

storage dynamics in a till hillslope: The foundation for modeling flows and residence times. *Hydrol. Process.* **2015**, submitted.

37. Grip, H.; Rodhe, A. *Vattnets väg från regn till back*; Hallgren & Fallgren: Uppsala, Sweden, 2000.

38. Bengtsson, L. Snowmelt estimated from energy budget studies. *Nord. Hydrol.* **1976**, *7*, 3–18.

39. Grip, H.; Halldin, S.; Lindroth, A. Water use by intensively cultivated willow using estimated stomatal parameter values. *Hydrol. Process.* **1989**, *3*, 51–63.

Chapter 6

COHERENT STRUCTURES IN FLOW OVER HYDRAULIC ENGINEERING SURFACES

Ronald J. Adrian

Ira A. Fulton Professor, School for Engineering of Matter, Transport and Energy, Arizona State University, Tempe, AZ 85287, USA

ABSTRACT

Wall-bounded turbulence manifests itself in a broad range of applications, not least in hydraulic systems. Here, we briefly review the significant advances over the past few decades in the fundamental study of wall turbulence over smooth and rough surfaces, with an emphasis on coherent structures and their role at high Reynolds numbers. We attempt to relate these findings to parallel efforts in the hydraulic engineering community and discuss the implications of coherent structures in important hydraulic phenomena.

INTRODUCTION

Flows over surfaces in hydraulic engineering are almost always intensely turbulent, owing to the low viscosity of water and the characteristically large scales of length, δ_0, and mean flow velocity, U. The archetypes for this class of flows are steady mean motions over smooth, flat surfaces with large fetch, for example, turbulent boundary layers or internal wall flows such as those in pipes and channels.

Classically, understanding of these flows is based largely on the average behaviour of the important aspects of the flow such as mean velocity and mean wall-shear stress, τ_w. The mean velocity exhibits at least in two different layers, an *inner layer* in which the wall-shear stress, expressed in terms of the friction velocity, $u_\tau = \sqrt{\tau_w/\rho}$, and the kinematic viscosity, ν, are the important external parameters; and an *outer layer* in which the depth of the flow δ_0 (equal to the boundary layer thickness δ, the pipe radius R or channel depth h) and the free stream velocity U_∞ or the bulk velocity U_b determine the average behaviour of the mean velocity profile. These layers share a common part, the *logarithmic layer*, in which the mean velocity varies logarithmically with distance from the wall, y. Coles' logarithmic plus wake formulation (Coles 1956) gives the mean velocity in the outer layer according to

$$U^+ \equiv \frac{U(y)}{u_\tau} = \kappa^{-1}\ln(y^+) + A + \Pi W(y/\delta_0), \quad y^+ > 30 \qquad (1)$$

Where von Kármán's constant, $\kappa \cong 0.41$ and $A \cong 5$ are empirical constants, and *Coles' wake factor* Π is an empirical, non-dimensional parameter that depends upon the free stream pressure gradient. The empirical fit $W \approx \sin^2(y/\delta_0)$ describes the deviation of the mean velocity from the logarithmic variation in the so-called *wake region*, and $y^+ = yu_\tau/\nu$ is the distance from the wall in units of the *viscous length scale*, ν/u_τ. The logarithmic variation dominates for $y \leq 0.15\delta_0$, nominally.

The mean velocity in the inner layer is described classically by von Kármán's logarithmic law above $y^+ \cong 30$, and a viscously dominated *buffer layer* for $0 \leq y^+ \leq 30$. (Modern investigations suggest that the mean velocity does not vary logarithmically until higher values, $y^+ \geq 200$ in boundary layers (Nagib *et al.* 2007) and 600 in pipes (Zagarola and Smits 1998), but for the purposes of this discussion, it suffices to use $y^+ = 30$ for reference.) Thus, the logarithmic layer nominally exists between

$$\frac{30}{R_\tau} < \frac{y}{\delta_0} < 0.15 \qquad (2)$$

Where

$$R_\tau \equiv \frac{u_\tau \delta_0}{\nu} = \delta_0^+ \qquad (3)$$

Can be interpreted either as a turbulent Reynolds number or as the ratio of the layer depth to the viscous length scale, known as the *von Kármán number*.

Neo-classically, there has been considerable research effort to understand the behaviour of the flow statistics in terms of structural elements, variously called motions, coherent structures or *eddies* (Townsend 1976, Cantwell 1981, Hussain 1986). Coherent motions are recurrent, persistent motions that characterize the flow and play important roles in determining mean flow, stress and other statistical properties. They may have rotational and irrotational parts. Eddies are similar, but in the spirit of Townsend (1976) they are definitely rotational. Further discussion can be found in Marusic and Adrian (2013), but for the present purposes it suffices to think of coherent structures as building blocks of flows that are recognizable, despite randomness, by their common topological patterns, and that occur over and over again.

The quantitative validity of the logarithmic variation of the mean velocity and the scaling laws that pertain to it have been questioned (Barenblatt 1993), especially for boundary layers (George and Castillo 1997), but there is now no doubt (Smits *et al.* 2011) that the logarithmic law continues to be one of the cornerstones of wall turbulence, and that the physics of the logarithmic region play a central role in the overall fluid mechanics of wall turbulence. This role extends to important issues such as the proper boundary conditions for Reynolds-averaged Navier–Stokes equations and large eddy simulations, and to the asymptotically infinite Reynolds number structure of the eddies of wall turbulence.

Despite the clear importance of the logarithmic layer at high Reynolds number and over a variety of surfaces, surprisingly little of our knowledge about the structures of eddies within the logarithmic layer is used in the treatment of hydraulic wall flows. For example, it is well known that the logarithmic law can be derived by postulating that the mixing length grows in proportion to y, and that it varies qualitatively as shown in Fig. 1(a). This proportionality in the logarithmic layer is consistent with Townsend's *Attached Eddy Hypothesis*, which states that the eddies in wall turbulence have sizes that are proportional to their distance from the wall (Fig. 1b). But, very little else about the geometry of the eddies, their origin or their dynamics is used in the classical hydraulic engineering literature.

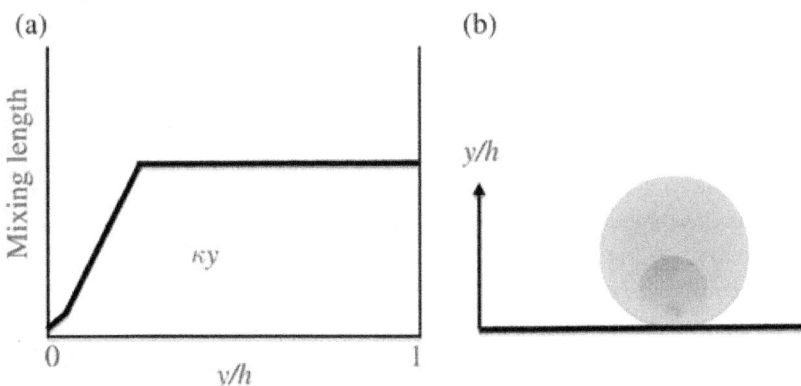

Figure 1. (a) Classical mixing length profile; (b) schematic illustration of Townsend's attached eddy hypothesis in which the attached eddies grow in size in proportion to their distance from the wall.

The place of understanding coherent structures within the hydraulics research portfolio is developing, and its ultimate applications remain to be established. Certainly, understanding how structures create motions that transport momentum, energy and scalars can be expected to materially improve the ability to predict average behaviour. Further, understanding the component structures of a turbulent flow is also likely to provide a conceptual framework within which observations of hydraulic phenomena can be assessed. Lastly, understanding the coherent structures may make the design of hydraulic structures easier.

The purpose of this "vision paper" is to summarize what is known about the structure of coherent structures in wall turbulence, especially the high Reynolds number turbulence of hydraulic flow applications, and to offer some ideas on the significance of the structures in problem areas such as sedimentation, erosion and flow–structure interactions. Throughout, we shall relate the coherent structures to the known regions of the mean velocity profile, as discussed above.

COHERENT STRUCTURES ON SMOOTH WALLS

Near-wall Structures

Before considering rough and irregular surfaces, it is valuable to consider the large body of work done on hydrodynamically smooth surfaces. Particularly, as theory (Townsend 1976, Jimenez 2004) indicates that for roughness length scales less than a few percent of the boundary layer thickness, the logarithmic and fully outer regions are not affected by roughness, apart from setting the inner boundary condition for the friction velocity, u_τ.

The coherent structures that occur in the near-wall portion of the inner layer have been extensively reviewed by Kline (1978), Cantwell (1981), Hussain (1986), Robinson (1991), Adrian (2007) and others. Many characteristic elements have been recognized and documented in the near-wall layer, including: low-speed streaks with spacing of 100 viscous wall units and the burst process (Kline *et al.* 1967), sweeps and ejections (Brodkey *et al.* 1974), quasi-streamwise vortices, Q2/Q4 events (Wallace *et al.* 1972, Willmarth and Lu 1972) and associated variable integration time average (VITA) events (Blackwelder and Kaplan 1976) and inclined shear layers (Kim 1987). Here, Q2/Q4 refers to events in the second and fourth quadrants of the u–v map, which thus contribute a positive contribution to the Reynolds shear stress, $-\overline{uv}$. It is noted here that we define u and v as the fluctuating components of velocity in the streamwise and wall-normal directions, respectively. The bursting process in the near-wall region, in which low-speed fluid is ejected abruptly away from the wall, is considered to play an important role in the overall dynamics of the boundary layer.

Different interpretations exist as to what type of coherent structures exist and what role they play in the near-wall region, and many of these viewpoints are reviewed by Robinson (1991), Panton (2001), Schoppa and Hussain (2002), Adrian (2007), Marusic *et al.* (2010b) and Jimenez (2012). Here, we emphasize the hairpin vortex as a simple coherent structure that explains many of the features observed in the near-wall layer (Theodorsen 1952, Head and Bandyopadhyay 1981), or it's more modern, and demonstrably more common variant, the asymmetric hairpin or the cane vortex (Guezennec *et al.* 1989, Robinson 1991, Carlier and Stanislas 2005). For brevity, we shall not distinguish between symmetric and asymmetric hairpins, nor will we distinguish between hairpins and horseshoes, since available evidence suggests that these structures are variations of a common basic structure at different stages of evolution or in different surrounding flow environments. In this regard, it may be also useful to group all such eddies into the class of *turbines propensii* (referring to "inclined eddies") to de-emphasize the connotations of shape that are intrinsic to the term "hairpin".

Theodorsen's (1952) analysis considered perturbations of the spanwise vortex lines of the mean flow that were stretched by the shear into intensified hairpin loops. Smith (1984) extended this model and reported hydrogen bubble visualizations of hairpin loops at low Reynolds number. While there is evidence for a formation mechanism like Theodorsen's in homogeneous shear flow (Rogers and Moin 1987, Adrian and Moin 1988), it is clear that Theodorsen's model requires modification near a wall to include long quasi-streamwise vortices spaced about 50 viscous wall units apart and connected to the head of the hairpin by vortex necks inclined at roughly 45° to the wall (Robinson 1991). With this simple model, the low-speed streaks are explained as the viscous sub-layer, low-speed fluid that is induced to move up from the wall by the quasi-streamwise vortices. A schematic illustrating these essential features of a hairpin vortex is shown in Fig. 2. The second quadrant ejections are the low-speed fluid that is caused to move through the inclined loop of the hairpin by vortex induction from the legs and the head, and the VITA event is the stagnation point flow that occurs when the Q2 flow through the hairpin loop encounters a Q4 sweep of higher speed fluid moving towards the back of the hairpin. This part of the flow constitutes the inclined shear layer. This picture is substantiated by the direct experimental observations of Liu *et al.* (1991), who used particle image velocimeter (PIV) to examine the structure of wall turbulence in the streamwise wall-normal plane of a fully developed low Reynolds number channel flow. They found shear layers growing up from the wall which were inclined at angles less than 45° from the wall. Regions containing high Reynolds stress were associated with these near-wall shear layers. Typically, these shear layers terminate in regions of rolled-up spanwise vorticity, which could be the heads of hairpin vortices. In the near-wall hairpin model, ejections are associated with the passage of hairpin vortices.

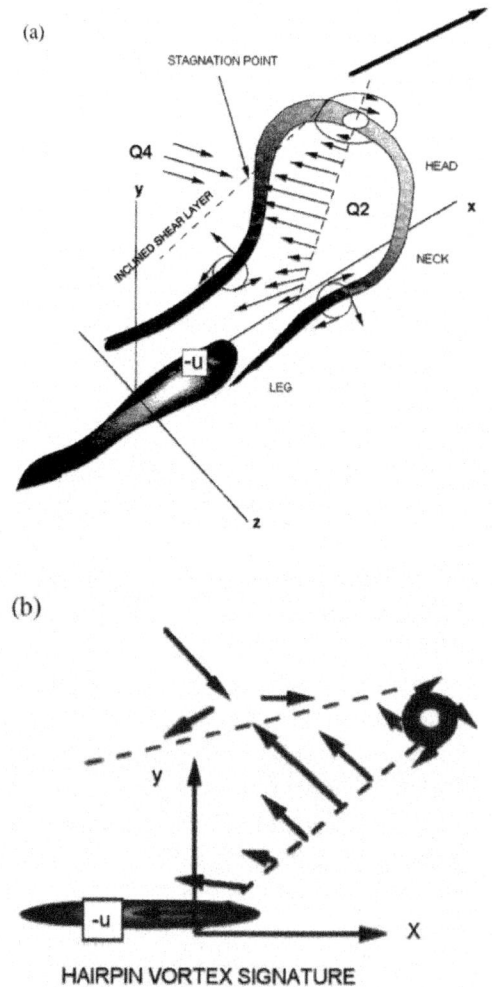

Figure 2. Schematic of hairpin eddy attached to the wall; (b) signature of the hairpin eddy in the streamwise/wall-normal plane (from Adrian *et al.* 2000).

Perhaps the strongest experimental support for the existence of hairpin vortices in the logarithmic layer was originally given by Head and Bandyopadhyay (1981), who studied high-speed, time-sequenced, images of smoke-filled boundary layers over a large Reynolds number range. They concluded that the turbulent boundary layer consists of hairpin structures that are inclined at a characteristic angle of 45° to the wall. Head and Bandyopadhyay (1981) also proposed that the hairpins occur in groups whose heads describe an envelope inclined at 15–20° with respect to the wall. The picture is similar to Smith's (1984)interpretation of flow visualizations in water, but instead of being based on data below $y^+=100$, Head and Bandyopadhyay(1981) appear to have based their construct on direct observations of ramp-like patterns on the *outer* edge of the boundary layer (Bandyopadhyay 1980), plus more inferential conclusions from data within the boundary layer. The observations of Head and Bandyopadhyay (1981) led Perry and Chong (1982), with later refinements by Perry *et al.* (1986) and Perry and Marusic (1995), to develop a mechanistic model for boundary layers based on Townsend's (1976) attached eddy hypothesis where the statistically representative attached eddies are hairpin vortices.

An important aspect of the attached eddy modelling work is that a logarithmic region requires a range of scales to exist with the individual eddies scaling with their distance from the wall. However, achieving such a range of scales requires a sufficiently high Reynolds number, which makes measurements difficult due to the large dynamic range required. A major advance in this regard came with the development of high-resolution PIV. Adrian *et al.* (2000) were the first to extensively use PIV to study the logarithmic and fully outer regions of boundary layers over a range of Reynolds numbers. Their work was particularly important as the PIV measurements provided images of the distribution of vorticity and the associated induced flow patterns without invoking the inferences needed to interpret flow visualization patterns. The patterns revealed that the logarithmic region is characterized by spatially coherent packets of hairpin vortices, with a range of scales of packets coexisting. This scenario explained the observed inclined regions of uniform momentum where the interfaces of these regions coincided with distinct vortex core signatures. A sample instantaneous PIV result is shown in Fig. 3. The "attached" hairpin packet scenario explains, or at least is consistent with a number of observations made in turbulent boundary layers. For example, it explains the observation that the spacing of the low-speed streaks in the streamwise velocity fields increases across the logarithmic region with distance from the wall (Ganapathisubramani *et al.* 2003, 2005, Tomkins and Adrian 2005). Moreover, if one associates a burst with a packet of hairpins, this construct offers an explanation both for the long extent of the near-wall low-speed streaks and for the occurrence of multiple ejections per burst, which has been documented in a number of studies (Bogard and Tiederman 1986, Luchik and Tiederman 1987, Tardu 1995). Thus, the original conception of a turbulent burst being a violent eruption in time is replaced by a succession of ejections due to the passage of a packet of hairpin vortices, the smallest hairpin creating the strongest ejection velocity.

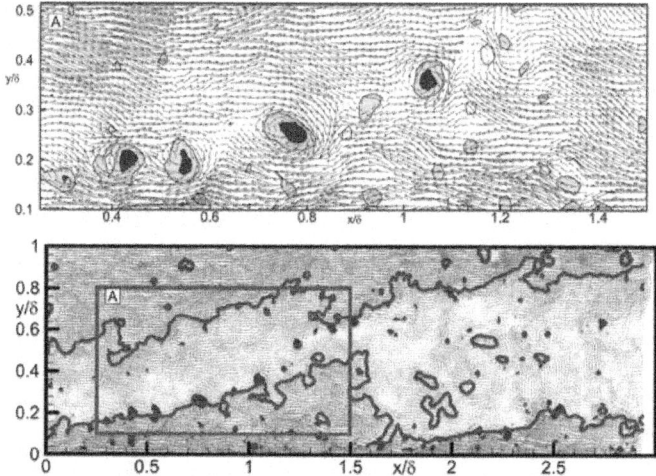

Figure 3. PIV measurements of the velocity field and vorticity field (coloured contours) in a turbulent boundary layer flowing left to right. The ramp-like structures bounded by groups of concentrated vorticies are evidence of hairpin vortex packets in which hairpins occur in a streamwise alignment with smaller, upstream hairpins auto-generated by larger, downstream hairpins. The velocity fields magnified in the upper inset figure possess the characteristics of hairpins identified in Figure 2 (from Adrian *et al.* 2000).

Large-scale Motions and Very Large-Scale Superstructures

Flow visualizations of boundary layers, an example of which is shown in Fig. 4, highlight that in the outer layer, the edge of the turbulent zone has bulges that are about 2–3 δ long (Kovasznay *et al.* 1970) separated by deep crevasses between the back of one bulge and the front of another (Cantwell 1981). The backs have stagnation points formed by high-speed fluid sweeping downward, and the shear between the high-speed sweep and the lower speed bulge creates an inclined, δ-scale shear layer. The bulges propagate at about 80–85% of the free stream velocity.

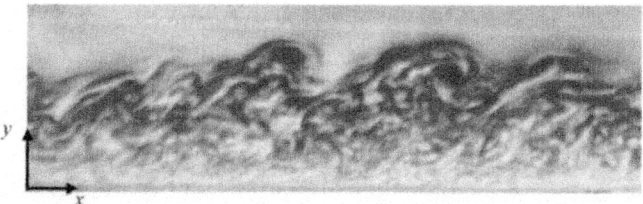

Figure 4. Flow visualization of a turbulent boundary layer. Flow is from left to right and the visualization details are as described in Cantwell *et al.* (1978). Photo courtesy of Don Coles.

Long streamwise lengths\enlargethispage are also prominent in streamwise velocity energy spectra, as reported by Balakumar and Adrian (2007). They showed that two large length scales emerge in pipe, channel and boundary layer flows where one peak in energy is associated with large-scale motions (LSMs) of typical length 2–3δ, and a second longer wavelength peak is associated with very-large-scale motions (VLSM), or superstructures, on the order of 6δ for boundary layers (Hutchins and Marusic 2007a). On the basis of the shapes of the streamwise power spectra and the uv co-spectra, Balakumar and Adrian (2007) nominally placed the dividing line between LSM and VLSM at 3δ. Using this demarcation, Balakumar and Adrian (2007) showed that the LSM wavelength persists out to about $y/\delta \sim 0.5$ (consistent with the observed bulges in visualizations), while the very large superstructure wavelengths do not extend beyond the logarithmic region, ending at approximately $y/\delta = 0.2$.

While the reported lengths for the very large superstructure events from spectra are approximately 6δ for boundary layers, this is considerably less than the observed values in pipe and channel flows (Kim and Adrian 1999, Monty *et al.* 2007, 2009), suggesting that geometrical confinement issues may play a role. However, what the actual lengths of the very large superstructures are remains an open question. Hutchins and Marusic (2007a) used time-series from a spanwise array of hot-wires (and sonic anemometers in the atmospheric surface layer) to infer lengths well in excess of 10 δ, and this is consistent with the high-speed PIV study of Dennis and Nickels (2008). Sample results of instantaneous measurements from Hutchins and Marusic (2007a) and Dennis and Nickels (2011a) are shown in Fig. 5.

The Dennis and Nickels (2011a, b) results also shed invaluable information on the three-dimensional structure of the largest motions, and while not conclusive, strongly support the suggestion by Kim and Adrian (1999) that the very large superstructures are a result of a concatenation of packets. Support for this also comes from atmospheric surface layer and laboratory measurements as described in Hambleton *et al.* (2006) and Hutchins *et al.* (2012), as shown in Fig. 6, where simultaneous x–y and x–z plane three-component velocity measurements reveal signatures entirely consistent with the superstructure events consisting of an organized array of packet structures. The lower schematics in Fig. 6 indicate comparisons with the Adrian *et al.* (2000) packet paradigm with Biot–Savart calculations of an idealized packet of hairpin vortices to infer what the corresponding spanwise velocity signatures would be in the relevant orthogonal planes.

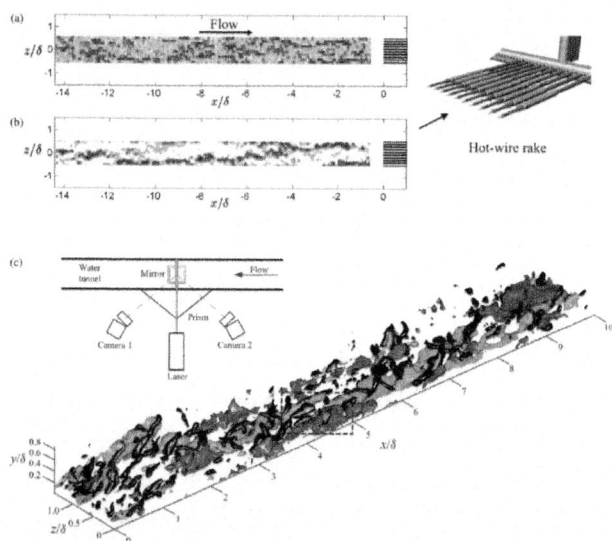

Figure 5. Very large-scale superstructure signatures: (a) from rake of hot-wire traces from Hutchins and Marusic(2007a); u signal at y/δ=0.15 for R_τ=14, 400. (b) Same with only low-speed regions highlighted. (c) High-frame rate stereo-PIV measurements from Dennis and Nickels (2011a, b) in a turbulent boundary layer at R_δ=4700, showing similar features to the hot-wire rake measurements. Here, the black isocontours show swirl strength, indicating the corresponding location of vortical structures with the low-speed (blue) and high-speed (red) regions. After Marusic and Adrian (2013).

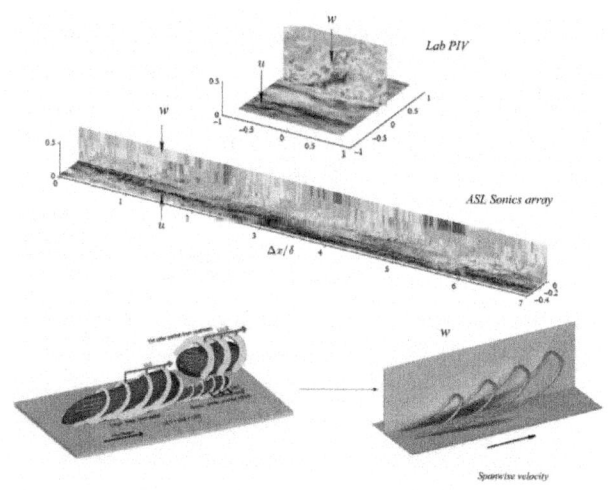

Figure 6. Top panel: instantaneous velocity fluctuations in the streamwise-wall-normal $(x–y)$ plane and instantaneous streamwise velocity fluctuations in the streamwise/

spanwise (x–y) planes for data from laboratory PIV (Hambleton *et al.* 2006) and for the atmospheric surface layer using an arrays of sonic anemometers (Hutchins *et al.* 2012). High-positive w regions are indicated by red, while blue denotes highly negative w regions. High-negative u regions are indicated by dark grey, while light grey shade denotes highly positive u regions. Bottom panel shows the Biot Savart law calculations for an idealized packet of hairpin vortices with their image vortices in the wall, as per the schematic of Adrian *et al.* (2000) shown on the left side.

Interactions across Scales

An important consequence of the large-scale and very large superstructure motions in the outer region (which includes the logarithmic region) is their role in interacting with the inner near-wall region, including their influence on the fluctuating wall-shear stress. There has been debate over many decades as to whether the inner and outer regions do interact, or whether they can be considered as independent, as assumed in all classical scaling approaches. Considerable evidence now exists that outer scales are important for characterizing near-wall events. This stems from a large number of studies that have documented a Reynolds number (or equivalently an outer length scale) dependence in the near-wall region. These include the studies of Rao *et al.* (1971), Blackwelder and Kovasznay (1972), Wark and Nagib (1991), Hunt and Morrison (2000), DeGraaff and Eaton (2000), Metzger and Klewicki (2001), Abe *et al.* (2004), Hoyas and Jimenez (2006), Hutchins and Marusic (2007a), Orlu and Schlatter (2011) and others. Many of the above studies support the viewpoint that some superposition of the LSMs is experienced right to the wall. Hutchins and Marusic (2007b) went further and proposed that this interaction also involved a modulation of the large scales on the near-wall small-scale motions. Previous suggestions of modulation effects have also been made by Grinvald and Nikora (1988). Mathis *et al.* (2009) studied the modulation effect extensively using data over a large range of Reynolds number and showed that the degree of modulation increased with increasing Reynolds number, and hence is a key aspect of high Reynolds number wall turbulence.

Marusic *et al.* (2010a) extended the observations of a superposition and modulation of the large-scale outer motions in the near-wall region to a predictive model, whereby a statistically representative fluctuating streamwise velocity signal near the wall could be predicted given only a large-scale velocity signature from the logarithmic region of the flow. The model was shown to work well over a large Reynolds number range for various statistics, including higher order moments. The formulation involves a universal signal and universal parameters, which are determined from a once-off calibration experiment at an arbitrarily chosen (but sufficiently high) Reynolds number.

Marusic *et al.* (2011) further extended the model to predict the fluctuating wall-shear stress given only a large-scale streamwise velocity signal from the logarithmic region, and were able to reproduce the empirical result of Alfredsson *et al.* (1988) and Orlu and Schlatter (2011) that showed that the standard deviation of the inner-scaled fluctuating wall-shear stress increases as a logarithmic function of the Reynolds number.

EFFECT OF HIGH *R* IN HYDRAULIC ENGINEERING

The significance of the logarithmic layer depends on the Reynolds number. At low Reynolds number, most of the change of the velocity from the wall to the free-stream occurs from the wall to the top of the viscous-inertial buffer layer because the thickness of the logarithmic layer is small, and there is relatively little change in the velocity in the wake region. For example, in turbulent channel flow at Reynolds number $R_\tau = 180$ (corresponding to $U_b h / v = 2800$), the mean velocity at the edge of the buffer layer is approximately 75% of the centreline velocity, and the velocity change across the logarithmic layer is very small. If one interprets the skin friction coefficient as a quantity that specifies the free stream velocity corresponding to a given level of wall shear stress, the foregoing consideration indicates that over half of the skin friction coefficient is determined by the fluid mechanics of the buffer layer at low Reynolds number, and hence that drag reduction strategies must concentrate on modifying the flow in the buffer layer. This view is supported by the fact that the rate of production of turbulent kinetic energy per unit volume, $-\overline{uv}\partial U/\partial y$, achieves a large maximum within the buffer layer, while it is much smaller in the logarithmic layer, suggesting that the preponderance of the turbulence is created in the buffer layer at low Reynolds number.

However, at high Reynolds numbers these conclusions must be altered substantially, simply because the logarithmic layer becomes much thicker, and thereby becomes more important. Consider for the sake of estimation equation (1). The velocity change from the wall to the top of the buffer layer is 13.2 friction velocities, while the velocity change from the top of the buffer layer to the top of the log layer (using $y/\delta_0 = 0.15$) is $2.41 \ln \delta^+ - 12.8$. The ratio of the velocity rise across the logarithmic layer to the velocity rise across the buffer layer is $0.183 \ln \delta^+ - 0.97$, implying that the velocity change across the buffer layer vanishes as $\approx 5.5/\ln \delta_0^+$ for large Reynolds number. Thus, as Reynolds number becomes infinite, essentially all of the velocity change occurs across the logarithmic layer, and hence all of the skin friction is associated with the logarithmic layer.

Practically, this conclusion is too strong, because the logarithmic dominance increases very slowly. For example, for 90% of the velocity change from the wall to the top of the logarithmic layer to occur across the logarithmic layer, the Kármán number must exceed 10^{23}, far above the value achieved by any terrestrial flow. On the other hand, for typical Reynolds number laboratory flows (say, $\delta^+ = 2000$), the velocity changes across the buffer layer, logarithmic layer and wake region are nominally 50, 25 and 25% of the free stream velocity, respectively. Thus, the logarithmic layer does not dominate laboratory flows, but its contribution is very substantial.

Similar conclusions can be drawn regarding the contribution that the logarithmic layer makes to the total production of the turbulent kinetic energy. For example, while the production per unit volume does peak in the buffer layer, the volume of the logarithmic layer is much greater, so the ratio of the production integrated over the logarithmic layer to the total production from within the buffer layer grows as $\ln y^+$ as Reynolds number approaches infinity. They are equal at approximately $\delta_0^+ = 35,000$.

Considerations such as the foregoing plus others have led Smits *et al.* (2011) to conclude that a reasonable criterion for wall turbulence to be considered high Reynolds number is $\delta_0^+ > 13,300$ for boundary layers and $\delta_0^+ > 50,000$ for pipe flow. These values are achieved commonly in hydraulic flows, so it is safe to assert *that nearly all hydraulic flows are high Reynolds number wall turbulence.* (For example, the turbulent Reynolds number of a boundary layer in a water flow with a free stream velocity of 2.5 m/s and a depth of 1 m is approximately 100,000.) This simple rule implies that hydraulic wall turbulence:

1. Possesses a clear range of logarithmic behaviour in the mean velocity profile and a clear range of k−5/3 behaviour in the inertial sub-range of the power spectrum of the streamwise velocity.

2. Has larger production of turbulent kinetic energy in the logarithmic layer than in the buffer layer

3. Possess a spectral peak at very long wavelengths that is distinct from the spectral peak corresponding to the inner layer motions.

With regard to the coherent structures, high Reynolds number implies ample room for eddies to grow from their initially small scales at the wall to the depth of the flow. The range of scales in the outer layer increases as $\delta^+/100$, if we take 100 viscous wall units as the representative height of the smallest first-generation hairpin and δ as the tallest coherent structure. If attention is confined to the self-similar structures in the logarithmic layer, the scale ratio is approximately $0.15\delta^+/100 = 150$ at $R_\tau = 100,000$, making room for at least seven doublings of the original height of the smallest hairpin

$(100 \times 2^7 = 12,800 < 15,000)$. This implies seven or more different uniform momentum zones across the logarithmic layer.

ROUGHNESS EFFECTS ON COHERENT STRUCTURE

The surfaces bounding hydraulic flows are seldom smooth, and the height of the roughness elements can easily exceed the thickness of the viscous buffer layer at the high Reynolds numbers of hydraulic flows. Roughness elements disrupt the flow within the buffer layer, and they may completely destroy it, replacing the effects of fluid viscosity with the effects of wall roughness and replacing the viscous length scale with the roughness element length scale, k (of course, a thin viscous sublayer is still attached to the surface of roughness elements, but its very small thickness makes it dynamically insignificant). A measure of the importance of the roughness elements is the non-dimensional roughness element height $k^+ = ku_\tau/v$. Small values of k^+ correspond to incomplete roughness, and large values correspond to complete or *fully developed* roughness. While roughness may destroy the viscous buffer layer, it appears to have much less effect on the logarithmic layer, other than shifting the effective slip velocity of the logarithmic layer with respect to the wall (Townsend 1976). The logarithmic law in Eq. (1) is, thus, replaced by

$$U^+ = \kappa^{-1} \ln y^+ + B(k^+) + \Pi W(y/\delta_0) \tag{4}$$

We shall refer to this phenomenon as *robustness of the logarithmic layer*. The persistence of the logarithmic layer implies that the under-lying structures, such as hairpins packets and related *turbines propensii* also persist. Their form need not be identical to the structures over smooth walls, but the evidence suggests that they are not very different (Hommema and Adrian 2003, Guala, *et al.* 2012). We, therefore, adopt, as a working hypothesis for now, the idea that the structures in the outer layer of turbulent flow over rough walls having roughness elements that are smaller than the logarithmic layer are similar to those occurring in the outer layer of turbulent flow over smooth walls.

If the roughness elements become a significant fraction of the logarithmic layer, they can severely disrupt the self-similar structures, and the logarithmic layer is replaced by different behaviour. A hint as to how this may happen is contained in the companion paper to this paper (Guala *et al.* 2012) in which tall hemispherical roughness elements are placed sparsely on an otherwise smooth surface. Measurements show two types of structures co-existing: hairpin packets from the smooth surface and hairpin packets from the individual hemispheres. The essential difference between the two types is that the latter grow at a steeper angle than the former and each of the latter packets is rooted to the hemisphere that generates it, much like wake vortices shed from a stationary cylinder. This behaviour hints at the effects that might be expected from rivets on the surfaces of marine vessels or very large roughness elements in streams and beds, such as large rocks.

COHERENT STRUCTURES AND HYDRAULIC PHENOMENA

Turbulent transport plays a critical role in heat and mass transfer at the free surface, mixing and dispersion, erosion and sedimentation, inlet conditions to hydraulic devices, interaction with vegetation and, of course, resistance to flow. As such, insights into the coherent structures that influence transport provide new ways of looking at each of these phenomena (Nezu2005, Nikora *et al.* 2007, Nikora 2010, Grant and Marusic 2011).

Coherent Structures in Canonical Open-Channel Flows

Here, we consider flow in straight, wide channels of depth h with smooth walls, unless otherwise stated. The most obvious coherent feature of open-channel flow is the boil phenomena (Yalin 1992, Nezu and Nakagawa 1993). These localized, intense upwellings occur one after another in streaks along the streamwise direction with a spacing of approximately $2h$ (Tamburrino and Gulliver 1999), which corresponds, to the large-scale motions (bulges) in turbulent boundary layers. The streaks of boils coincide with streaks of low-speed flow, upwelling and lateral spreading at the surface. They are separated by streaks of high-speed flow lateral convergence and downwelling (Tamburrino and Gulliver 1999, 2007). From the upwelling and downwelling long, streamwise-oriented rolling vortices apparently first inferred by Velikaniv (1958; Shvidchenko and Pender 2001) and observed by many subsequent workers (Klaven and Kopaliani 1973 and more recently Tamburrino and Gulliver (1999, 2007), and Rodriguez and Garcia (2008) to cite a few).

The roll cells, also called large streamwise vortices (Gulliver and Halverson 1987) or long longitudinal eddies (Imamoto and Ishigaki 1986), look like secondary flows in the plane perpendicular to the streamwise flow (Nikora and Roy 2012). True secondary flows have non-zero long-time averages, and they affect the distribution of mean velocity, turbulence intensities, Reynolds shear stresses and bed shear stress throughout the channel. If the channel is wide enough, width $> 5h$ Nezu and Rodi (1986) observed that secondary flows are hard to see in the long-time averages, but they exist, nonetheless. PIV measurements of the cross-stream flow find cellular secondary currents that vary in time regardless of the aspect ratio (Onitsuka and Nezu 2001). This suggests that the long streamwise vortices meander in time as the aspect ratio increases, causing their features to be lost in time average measurements. Tamburrino and Gulliver (2007) observed that large-scale eddies having spanwise (lateral) widths of 1–1.5 h oscillate slowly in the mid channel, but fixed stationary secondary flows form in the vicinity of the side walls. Nezu and Nakayama (1997) observe both secondary currents and time-varying cellular currents in the interaction between the mainstream and a flood plain. Correlation measurements of the streamwise surface velocity made in many rivers indicate positive correlation over 2–5h followed by negative correlation between 5 and 10h, and finite correlation, either positive or negative over lengths extending to 10–20h (Sukhodolov *et al.* 2011). The oscillating sign of the correlation in Sukhodolov *et al.* (2011) implies that the streaks either waver or drift laterally so that a streamwise line of observation alternately crosses high-speed and low-speed streaks.

A simple drawing summarizing these features is presented in Fig. 7. Note that the secondary flows are steady and aligned with the side-walls, and the long streamwise vortices are unsteady and inclined. While the cellular picture in Fig. 7 is appealing, the reality of open-channel flows is more complicated. Direct observations of multiple circulations perpendicular to the main channel flow have been made by Nezu (2005), and their instantaneous streamlines clearly fluctuate considerably from cell to cell. Furthermore, the cells do not appear to extend down to the bed. Consequently, the interior cells in Fig. 7 are too regular to represent the instantaneous flow, and the reader should think of them as a conditional average of the roll cells given the location of the centre of the cell as it meanders.

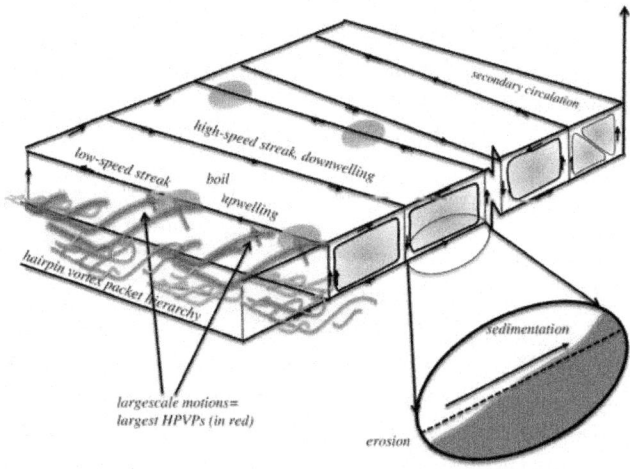

Figure 7: Cartoon of coherent structures in open-channel flows.

The irregularity of real roll cells can be explained in part by their close association with turbulent "bursts" in the low-speed zones. The term "burst" will be used in the present discussion in deference to common usage in the hydraulics literature. However, there is good evidence that the concept of a burst as a rapid, perhaps even violent, ejection should be replaced by the concept of a packet of hairpin vortices passing and creating a sequence of ejection events, each associated with one of the hairpins. Since the packet evolves relatively slowly, the appearance of rapid change is caused by the fast passage of the packet (Adrian *et al.* 2000). Observations show that a burst can originate at the bed and cross the entire channel depth to impinge on the surface and cause a boil (cf. Shvidchenko and Pender 2001 for a summary of the observations). The bursts reaching the surface have height h, length $2–5h$ and width $1–2h$, virtually the same as the large-scale motions or bulges discussed earlier. In turbulent boundary layers, the bulges are likely to be the ultimate form assumed by the hairpin vortex packets upon reaching the edge of the boundary layer. Consequently, Fig. 7 indicates hairpin vortex packets of various sizes, with the largest (coloured red) causing the surface boils. The smaller packets grow and merge with others to ultimately form the largest packets. PIV measurements in the streamwise vertical plane strongly support the similarity between internal packets in open-channel flow and turbulent boundary layers (Nezu and Sanjou 2011, Fig. 5).

While the association between the low-speed streaks and the succession of bursts that creates "street" of boils is well established, there is a very interesting issue of cause and effect. Shvidchenko and Pender (2001) assert that bursts give rise to the long, streamwise-oriented rolling vortices. But, in their reply to this discussion, Tamburinno and Gulliver note that the rolling vortices may *cause* the ejections and the sweeps, rather than vice versa. A similar idea has been developed independently in the turbulence community. The evidence presented earlier for modulation of the small near-wall scales by the large outer scales supports this picture. The authors' view is that both mechanisms are plausible, and that it is likely that they operate cooperatively. In this scenario, the lateral motion of the cells towards the low-speed streaks sweeps the smaller, growing hairpins and packets into the streaks (Toh and Itano 2005, Adrian 2007) and create the alignment of the LSMs. That alignment creates the VLSMs. Since the hairpins and packets are themselves elements of low momentum, their congregation around the VLSM's low-speed streaks intensifies the momentum deficit. Schoppa and Hussain (2002) have shown that low-speed streaks are necessarily associated with quasi-streamwise roll cells, so intensified low momentum would actually support formation of the roll cells. In this way, a closed-loop feedback cycle would exist in which the roll cells feed themselves by sweeping low-momentum hairpins and packets into the low-speed streaks. The close relationship between the meandering VLSMs of turbulence structure research and the long cellular motions of open-channel flow research is impossible to ignore. It seems likely, in fact that they are one and the same. Sukhodolov *et al.*(their Fig. 5b), shows correlation out to 5–10*h* in a compilation of time delayed streamwise correlation functions from many rivers, and their Fig. 5(a) shows alternating high-speed low-speed zones extending up to free surface. This is very similar to results for meandering VLSMs in pipes, channels and turbulent boundary layers and the atmospheric boundary layer.

Structure in Channels with Significant Roughness

Understanding of coherent structures in rough walled channels is limited, but generally speaking the picture is similar to that for smooth walls, as described in the detailed study by Detert *et al.* (2010). Several observations report structures resembling LSMs that grow up from the wall and reach the surface (Roy *et al.* 2004, Hurther *et al.* 2007, Nikora *et al.* 2007). Surface lengths of 3–5*h* are reported, but observed widths of 1*h* are somewhat smaller than the 1–1.5*h* width of turbulent bulges. It is well known that rough walls reduce the streamwise correlation length. Flow visualization of the bursts from the bed (Roy *et al.*2004, Fig. 16) shows structures whose growth angle looks similar to the $\sim 15°$ angle of hairpin vortex packets, followed by structures that grow much more rapidly, at least 45°. The latter probably emanates from single roughness elements, and the rapid growth angle offers the simplest explanation for the foreshortening of the streamwise length. The companion paper by Guala *et al.* (2012) offers some insight into the structures created by sparse roughness elements.

Heat and Mass Transfer at the Free Surface

Free surface boils and other structures at the surface are hydraulic manifestations of coherent structure rising to the surface. The interactions of the coherent structures with the free surface are also important in the gas exchange at the surface, a major factor in evaluation of greenhouse gas effects. The boils and the upwelling/downwelling streaks are the basis for surface renewal theories, as discussed by Komori *et al.* (1982). In this regard, Calmet and Magnaudet (2003) have shown the significance and utility of Hunt and Graham's (1978) rapid distortion theory for eddies approaching a surface, and this looks like a promising improvement on surface renewal theory.

Mixing and Dispersion

Mixing is perhaps one of the most important turbulent processes in problems involving dilution of thermal and material effluents and density stratification in hydraulic flows. Since the importance of coherent structures in the transport of momentum has been established conclusively, it is clear that the transport of heat and mass must also exhibit a strong dependence upon coherent structures. The dispersion of heat and pollutants may be affected by the structure of wall turbulence in shallow channel flows. Jirka (2001) studied wakes, jets and shear layer in wide open channels and noted that three-dimensional turbulent bursts can affect these mainly two-dimensional flows.

Dispersion of scalars is classically modelled as a random walk process that occurs on top of a mean flow field (Sawford 2001, Balachandar and Eaton 2010). The random walk naturally leads to concentration fields caused by dispersion from a point source that are Gaussian functions of position. But in reality, the coherent structures in the flow produce a different picture of the dispersion process. The anisotropy of the structures and their inhomogeneity are factors that are difficult to incorporate realistically into Gaussian models, and the short-term inhomogeneity that is associated with very large-scale superstructures, and their associated large streaks, is almost never accounted for. If the surface were flat and wide, the long streaks would meander with no preferred spanwise location, so that long-time averages would indeed be independent of the spanwise location. But over short times, the streaks tend to stay in one location, causing substantial inhomogeneity. The presence of small-scale inhomogeneity such as rocks, asperities, etc. could cause the streaks to stabilize, meaning that spanwise inhomogeneity would be lost. In such cases, it is very important to model the realizations of the coherent structures rather than their long-time mean values.

Erosion and Sedimentation

Erosion and sedimentation often lead to the formation patterns in solid boundaries such as dunes and meanders in streams, and it is, therefore, not unreasonable to look for associations between the formation of these patterns and the coherent patterns of flow in the fluid, at least in the incipient or early stages of erosion when the bed form is essentially flat. Gyr and Schmid (1997) have shown that at incipient erosion on a flat sandy bed, only the sweeps move the sand gains. Erosion processes are also likely to feel the consequences of coherent structures because the low probability, extreme events responsible for high-local erosion rates are parts of the natural cycle of flow. Roughness can also create fluctuation in the wall shear stress that are comparable to the fluctuations caused by coherent structures in smooth-walled flows (Cheng

2006). When sedimentation and erosion are strong enough to alter the bed form, the coherent structures above the bed may be radically modified, especially by the process of flow separation. For example, Kadota and Nezu (1999) show that the flow behind the crest of a dune is a turbulent shear layer containing spanwise vortices. Nezu *et al.* (1988) and Nezu and Nakagawa (1989b) found that the organized fluid motions and the associated sediment transport occurred intermittently on a movable plane sand bed. After the sand ridges were formed, the roll cells appeared stably across the whole channel cross section. As shown in the inset of Fig. 7, the sand is eroded in the downwelling side of a cell and sedimented on the upwelling side, roll cells are also generated on beds with smooth and rough striping (Nakagawa *et al.* 1981, McLean 1981, Studerus 1982). There is an extensive literature on the modification of turbulence statistics by various bed form geometries cf. Cellino and Graf (2000). A further comprehensive discussion of coherent structures in sediment dynamics can be found in Garcia (2008).

FUTURE CHALLENGES AND PROSPECTS

Our present knowledge of coherent structures in flows over smooth flat surfaces is enough to see how such structures could be of importance in hydraulic engineering. Efforts are needed to exploit understanding of the structure to improve hydraulic engineering design in many areas. Sedimentation, erosion, dispersion and entrance flows to hydraulic devices such as power facilities, spillways and barbs are importantly related to the large-scale and VLSMs, and considerable advancement can be expected if we can adequately characterize and predict these motions and possibly manipulate them in a controlled way. The interactions of the LSMs with the near-wall region, and thus the bed shear stress, also need to be studied and better exploited. Existing predictive models based on the outer region LSMs (Marusic *et al.* 2010b) need to be extended beyond smooth-wall flows and offer the prospect of real predictive capability given only the large-flow field information. Such information can be obtained by reasonably spatially-sparse, low-frequency measurements or preferably from numerical simulations, such as large eddy simulations, where the large flow field information is resolved. Fully understanding the scaling behaviour at high Reynolds numbers also opens the way for refined scale up from models and better-informed designs.

At this point in time the various types of structure have been identified, but one cannot claim that we fully understand their scaling or their functions. Investigations of the scaling of each type of motion are needed. They may provide better definitions of the motions and improve understanding of their relative importance in different ranges of Reynolds number. The interactions of the various motions have only begun to be understood, and much work, especially dynamic experiments and theoretical analyses are needed to establish true cause and effect in these interactions. For example, erosion by VLSMs may be caused by direct action of the VLSMs, but it may also be the case that the very large scales mainly organize and collect the smaller motions, and it is the latter that perform most of the erosion. Understanding cause and effect is essential to management of fluid flows by design.

It would be truly disappointing if improved understanding of the structures in turbulent flows and their roles in sedimentation, erosion and dispersion could not significantly improve the accuracy and reliability of turbulence models of all kinds. Ultimately, incorporation of structural properties into the models is one of the more important and more challenging tasks ahead of the field. It is hoped that improved paradigms of turbulent flow will stimulate new and innovative theoretical descriptions and computational modelling.

The very large Reynolds number inherent to hydraulic flows make them attractive for the study of turbulent structure in the presence of a wide hierarchy of scales and important to turbulent flow science. The wide range of scales across the logarithmic layer would be especially helpful in this regard. The persistence of the logarithmic layer and attached eddies above rough surfaces must be confirmed more fully, as this is an important piece of evidence concerning the robust nature of structures in the outer region. Acquiring such information experimentally will require resolving these flows with an unprecedentedly large dynamic range. However, rapid advances in laser and digital camera technologies combined with evolving three-dimensional velocimetry techniques (Adrian and Westerweel 2011) make this a realistic proposition in the not too distant future.

Acknowledgements

The authors gratefully acknowledge the support of NSF Grant CBET-0933848 (RJA) and the Australian Research Council (IM).

REFERENCES

1. Abe, H., Kawamura, H. and Choi, H. 2004. Very large-scale structures and their effects on the wall shear–stress fluctuations in a turbulent channel flow up to $R_\tau = 640$. *J. Fluids Eng.*, 126(9): 835–843. ,

2. Adrian, R. J. 2007. Hairpin vortex organization in wall turbulence. *Phys. Fluids*, 19(4): 1–16. 041301 ,

3. Adrian, R. J., Meinhart, C. D. and Tomkins, C. D. 2000. Vortex organization in the outer region of the turbulent boundary layer. *J. Fluid*

Mech., 422: 1–54. ,

4. Adrian, R. J. and Moin, P. 1988. Stochastic estimation of organized turbulent structure: Homogeneous shear flow. *J. Fluid Mech.*, 190: 531–559. , ,

5. Adrian, R. J. and Westerweel, J. 2011. *Particle image velocimetry*, Cambridge UK: Cambridge University Press.

6. Alfredsson, P. H., Johansson, A. V., Haritonidis, J. H. and Eckelmann, H. 1988. The fluctuating wall-shear stress and the velocity-field in the viscous sublayer. *Phys. Fluids*, 31(5): 1026–1033. , ,

7. Balachandar, S. and Eaton, J. K. 2010. Turbulent dispersed multiphase flow. *Ann. Rev. Fluid Mech.*, 42: 111–133. ,

8. Balakumar, B. J. and Adrian, R. J. 2007. Large- and very-large-scale motions in channel and boundary-layer flows. *Philos. Trans. R. Soc. Lond. A*, 365: 665–681. , ,

9. Bandyopadhyay, P. R. 1980. Large structure with a characteristic upstream interface in turbulent boundary layers. *Phys. Fluids*, 23: 2326–1980. ,

10. Barenblatt, G. I. 1993. Scaling laws for fully developed turbulent shear flows. Part 1. Basic hypotheses and analysis. *J. Fluid Mech.*, 248: 513–520. ,

11. Blackwelder, R. F. and Kaplan, R. E. 1976. On the wall structure of the turbulent boundary layer. *J. Fluid Mech.*, 76: 89–112.,

12. Blackwelder, R. F. and Kovasznay, L. S.G. 1972. Time scales and correlations in a turbulent boundary layer. *Phys. Fluids*, 15: 1545–1554. ,

13. Bogard, D. G. and Tiederman, W. G. 1986. Burst detection with single-point velocity measurements. *J. Fluid Mech.*, 162: 389–413. , ,

14. Brodkey, R. S., Wallace, J. M. and Eckelmann, H. 1974. Some properties of truncated turbulence signals in bounded shear flows. *J. Fluid Mech.*, 63: 209–224. ,

15. Calmet, I. and Magnaudet, J. 2003. Statistical structure of high-Reynolds-number turbulence close to the free surface of an open-channel flow. *J. Fluid Mech*, 474: 355–378. , ,

16. Cantwell, B. J. 1981. Organized motion in turbulent flow. *Ann. Rev. Fluid Mech.*, 13: 457–515. ,

17. Carlier, J. and Stanislas, M. 2005. Experimental study of eddy structures in a turbulent boundary layer using particle image velocimetry. *J. Fluid Mech.*, 535: 143–188. ,

18. Cellino, M. and Graf, W. H. 2000. Experiments on suspension flow in

open channels with bed forms. *J. Hydraulic Res.*, 38(4): 289–298. , ,

19. Cheng, N. 2006. Influence of shear stress fluctuation on bed particle mobility. *Phys. Fluids*, 18(9): 1–7. 096602 ,

20. Coles, D. E. 1956. The law of the wake in the turbulent boundary layer. *J. Fluid Mech.*, 1: 191–226. ,

21. Dennis, D. J.C. and Nickels, T. B. 2008. On the limitations of taylor's hypothesis in constructing long structures in a turbulent boundary layer. *J. Fluid Mech.*, 614: 197–206. ,

22. Dennis, D. J.C. and Nickels, T. B. 2011a. Experimental measurement of large-scale three-dimensional structures in a turbulent boundary layer. Part 1: Vortex packets. *J. Fluid Mech.*, 673: 180–217. ,

23. Dennis, D. J.C. and Nickels, T. B. 2011b. Experimental measurement of large scale three-dimensional structures in a turbulent boundary layer. Part 2: Long structures. *J. Fluid Mech.*, 673: 218–244. ,

24. DeGraaff, D. B. and Eaton, J. K. 2000. Reynolds-number scaling of the flat-plate turbulent boundary layer. *J. Fluid Mech.*, 422: 319–346. ,

25. Detert, M., Nikora, V. and Jirka, G. H. 2010. Synoptic velocity and pressure fields at the water–sediment interface of streambeds. *J. Fluid Mech.*, 660: 55–86. ,

26. Ganapathisubramani, B., Hutchins, N., Hambleton, W. T., Longmire, E. K. and Marusic, I. 2005. Investigation of large-scale coherence in a turbulent boundary layer using two-point correlations. *J. Fluid Mech.*, 524: 57–80. ,

27. Ganapathisubramani, B., Longmire, E. K. and Marusic, I. 2003. Characteristics of vortex packets in turbulent boundary layers. *J. Fluid Mech.*, 478: 35–46. , ,

28. Garcia, G. (2008). Sediment transport and morphodynamics. In Sedimentation engineering: Processes, measurements, modeling, and practice, 21–164, M. Garcia, ed. American Society of Civil Engineers, Manuals and Reports on Engineering Practice 110. Reston, Virginia.

29. George, W. K. and Castillo, L. 1997. Zero-pressure-gradient turbulent boundary layer. *App. Mech. Rev.*, 50(12): 689–729.

30. Grant, S. B. and Marusic, I. 2011. Crossing turbulent boundaries: Interfacial flux in environmental flows. *Environ. Sci. Technol.*, 45(17): 7107–7113. , ,

31. Grinvald, D. and Nikora, V. 1988. *River Turbulence (Rechnaya turbulentnosti) (in Russian)*, Russia>: Hydrometeoizdat.

32. Guala, M., Tomkins, C. D., Christiansen, K. T. and Adrian, R. J. 2012.

Vortex organization in turbulent boundary layer flow over sparse roughness elements. *J. Hydraulic Res.*, 50(5): 465–481. ,

33. Gulliver, J. S. and Halverson, M. J. 1987. Measurements of large streamwise vortices in an open-channel flow. *Water Resour. Res.*, 23(1): 115–123. , ,

34. Guezennec, Y. G., Piomelli, U. and Kim, J. 1989. On the shape and dynamics of wall structures in turbulent channel flow.*Phys Fluids A*, 1(4): 764–766. ,

35. Gyr, A. and Schmid, A. 1997. Turbulent flows over smooth erodible sand beds in flumes. *J. Hydraulic Res.*, 35: 525–544., ,

36. Hambleton, W. T., Hutchins, N. and Marusic, I. 2006. Simultaneous orthogonal-plane particle image velocimetry measurements in a turbulent boundary layer. *J. Fluid Mech.*, 560: 53–64. ,

37. Head, M. R. and Bandyopadhyay, P. R. 1981. New aspects of turbulent boundary-layer structure. *J. Fluid Mech.*, 107: 297–337. , ,

38. Hommema, S. E. and Adrian, R. J. 2003. Packet structure of surface eddies in the atmospheric boundary layer. *Boundary-Layer Meterorol.*, 106: 147–170. , ,

39. Hoyas, S. and Jimenez, J. 2006. Scaling of the velocity fluctuations in turbulent channels up to $R_\tau = 2003$. *Phys. Fluids*, 18(1): 1–4. 011702 ,

40. Hunt, J. C.R. and Graham, J. M.R. 1978. Free stream turbulence near plane boundaries. *J. Fluid Mech.*, 84: 209–235.,

41. Hunt, J. C.R. and Morrison, J. F. 2000. Eddy structure in turbulent boundary layers. *Eur. J. Mech. B-Fluids*, 19: 673–694., ,

42. Hurther, D., Lemmin, U. and Terray, E. A. 2007. Turbulent transport in the outer region of rough-wall open-channel flows: The contribution of large coherent shear stress structures (LC3S). *J. Fluid Mech.*, 574: 465–493. ,

43. Hutchins, N. and Marusic, I. 2007a. Evidence of very long meandering streamwise structures in the logarithmic region of turbulent boundary layers. *J. Fluid Mech.*, 579: 1–28. ,

44. Hutchins, N. and Marusic, I. 2007b. Large-scale influences in near-wall turbulence. *Philos. Trans. R. Soc. Lond. A*, 365: 647–664. , ,

45. Hutchins, N., Chauhan, K., Marusic, I., Monty, J. and Klewicki, J. 2012, in press. "Towards reconciling the large-scale structure of turbulent boundary layers in the atmosphere and laboratory". In *Boundary-Layer Meterol*

46. Hussain, A. K.M.F. 1986. Coherent structures and turbulence. *J. Fluid Mech.*, 173: 303–356. ,

47. Imamoto, H. and Ishigaki, T. Visualization of longitudinal eddies in an open-channel flow. Proc. 4th Int. Symp. Flow Visualization. 333–337. Edited by: Veret, C. Washington, DC: Hemisphere.

48. Jimenez, J. 2004. Turbulent flows over rough walls. *Annu. Rev. Fluid Mech.*, 36: 173–96. ,

49. Jimenez, J. 2012. Cascades in wall-bounded turbulence. *Ann. Rev. Fluid Mech.*, 44: 27–45. ,

50. Jirka, G. H. 2001. Large scale flow structures and mixing processes in shallow flows. *J. Hydraulic Res.*, 39(6): 567–573. , ,

51. Kadota, A. and Nezu, I. 1999. Three-dimensional structure of space-time correlation on coherent vortices generated behind dune crest. *J. Hydraulic Res.*, 37(1): 59–80. , ,

52. Kim, J. 1987. "Evolution of a vortical structure associated with the bursting event in a channel flow". In *Turbulent shear flows*, Edited by: Durst, F., Launder, B. E., Lumley, J. L., Schmidt, F. W. and Whitelaw, J. H. Vol. 5, 221–227. New York: Springer-Verlag.

53. Kim, K. C. and Adrian, R. J. 1999. Very large-scale motion in the outer layer. *Phys. Fluids*, 11(2): 417–422. ,

54. Klaven, A. B. and Kopaliani, Z. D. 1973. *Laboratory investigations of kinematic structure of turbulent flow over a very rough bed (in Russian)* Tarns. State Hydro. Inst. Gidrometeoizdat, Leningrad, Russia.

55. Kline, S. J. 1978. "The role of visualization in the study of the structure of the turbulent boundary layer". In *Lehigh workshop on coherent structure of turbulent boundary layers*, Edited by: Smith, C. R. and Abbott, D. E. 1–26. USA: Lehigh University.

56. Kline, S. J., Reynolds, W. C., Schraub, F. A. and Rundstadler, P. W. 1967. The structure of turbulent boundary layers. *J. Fluid Mech.*, 30: 741–773. ,

57. Komori, S., Ueda, H., Ogino, F. and Mizushina, T. 1982. Turbulence structure and transport mechanism at the free surface in an open channel flow. *Int. J. Heat Mass Trans.*, 25(4): 513–521. , ,

58. Kovasznay, L. S.G., Kibens, V. and Blackwelder, R. F. 1970. Large scale motion in the intermittent region of a turbulent boundary layer. *J. Fluid Mech.*, 41: 283–325. ,

59. Liu, Z. C., Adrian, R. J. and Hanratty, T. J. 1991. High resolution measurement of turbulent structure in a channel with particle image velocimetry. *Exp. Fluids*, 10(6): 301–312. ,

60. Luchik, T. S. and Tiederman, W. G. 1987. Time-scale and structure of ejections and bursts in turbulent channel flows. *J. Fluid Mech.*, 174: 529–552. , ,

61. Marusic, I. and Adrian, R. J. 2013. "Eddies and scales of wall turbulence". In *Ten chapters in turbulence*, Edited by: Davidson, P. A., Kaneda, Y. and Sreenivasan, K. R. Cambridge UK: Cambridge University Press.

62. Marusic, I., Mathis, R. and Hutchins, N. 2010a. Predictive model for wall-bounded turbulent flow. *Science*, 329(5988): 193–196. , ,

63. Marusic, I., Mathis, R. and Hutchins, N. 2011. A wall-shear stress predictive model. *J. Phys. Conf. Ser.*, 318(012003): 1–8.

64. Marusic, I., McKeon, B. J., Monkewitz, P. A., Nagib, H. M., Smits, A. J. and Sreenivasan, K. R. 2010b. Wall-bounded turbulent flows: Recent advances and key issues. *Phys. Fluids*, 22(6): 1–24. 065103 ,

65. Mathis, R., Hutchins, N. and Marusic, I. 2009. Large-scale amplitude modulation of the small-scale structures in turbulent boundary layers. *J. Fluid Mech.*, 628: 311–337. ,

66. Metzger, M. M. and Klewicki, J. C. 2001. A comparative study of near-wall turbulence in high and low Reynolds number boundary layers. *Phys. Fluids*, 13(3): 692–701. , ,

67. Monty, J. P., Hutchins, N., Ng, H. C.H., Marusic, I. and Chong, M. S. 2009. A comparison of turbulent pipe, channel and boundary layer flows. *J. Fluid Mech.*, 632: 431–442. ,

68. Monty, J. P., Stewart, J. A., Williams, R. C. and Chong, M. S. 2007. Large-scale features in turbulent pipe and channel flows. *J. Fluid Mech.*, 589: 147–156. ,

69. Nagib, H. M., Chauhan, K. A. and Monkewitz, P. A. 2007. Approach to an asymptotic state for zero pressure gradient turbulent boundary layers. *Philos. Trans. R. Soc. Lond. A.*, 365: 755–770. ,

70. Nakagawa, H., Nezu, I. and Tominaga, A. 1981. "Turbulent structure with and without cellular secondary currents over various bed configurations". In *Annu, DPRI*, Vol. 24B, 315–338. Kyoto Univ. in Japanese

71. Nezu, I. 2005. Open-channel flow turbulence and its research prospect in the 21st century. *J. Hydraulic Eng.*, 131(4): 229–246. ,

72. Nezu, I. and Nakagawa, H. 1993. *Turbulence in open-channel flows*, Balkema, Rotterdam The Netherlands: IAHR-Monograph.

73. Nezu, I. and Nakayama, T. 1997. Space-time correlation structures of horizontal coherent vortices in compound open-channel flows by using particle tracking velocimetry. *J. Hydraulic Res.*, 35(2): 191–208. , ,

74. Nezu, I., Nakagawa, H. and Kawashima, N. Cellular secondary currents and sand ribbons in fluvial channel flows. Proc. 6th APDIAHR Congress. Vol. 1, pp.51–58. Delft The Netherlands

75. Nezu, I. and Nakagawa, H. Self forming mechanism of longitudinal sand ridges and troughs. Proc. 23rd IAHR Congress. Vol. B, pp.65–72. Delft The Netherlands: IAHR.

76. Nezu, I. and Rodi, W. 1986. Open-channel flow measurements with a laser Doppler anemometer. *J. Hydraulic Eng.*, 112: 335–355. ,

77. Nezu, I. and Sanjou, M. 2011. PIV and PTV measurements in hydro-sciences with focus on turbulent open-channel flows. *J. Hydro-environment Res.*, 5(4): 215–230. ,

78. Nikora, V., Nokes, R., Veale, W., Davidson, M. and Jirka, G. H. 2007. Large-scale turbulent structure of uniform shallow free-surface flows. *Environ. Fluid Mech.*, 7(2): 159–172. ,

79. Nikora, V. 2010. Hydrodynamics of aquatic ecosystems: An interface between ecology, biomechanics and environmental fluid mechanics. *River Res. Applic.*, 26(4): 367–384. ,

80. Nikora, V. and Roy, A. G. 2012. "Secondary flows in rivers: Theoretical framework, recent advances, and current challenges". In *Gravel-bed Rivers: Processes, Tools, Environments*, Edited by: Church, M., Biron, P. M. and Roy, A. G. 3–22. USA: John Wiley & Sons.

81. Onitsuka, K. and Nezu, I. 2001. "Generation mechanism of turbulence driven secondary currents in open-channel flows. IUTAM Symp". In *Geometry and statistics of turbulence*, Edited by: Kambe, K., Nakano, T. and Miyauchi, T. 345–350. Boston: Kluwer Academic.

82. Orlu, R. and Schlatter, P. 2011. On the fluctuating wall-shear stress in zero-pressure-gradient turbulent boundary layers.*Phys. Fluids*, 23(2): 1–4. 021704

83. Panton, R. L. 2001. Overview of the self-sustaining mechanisms of wall turbulence. *Prog. Aerosp. Sci.*, 37: 341–383., ,

84. Perry, A. E. and Chong, M. S. 1982. On the mechanism of wall turbulence. *J. Fluid Mech.*, 119: 173–217. , ,

85. Perry, A. E., Henbest, S. M. and Chong, M. S. 1986. A theoretical and experimental study of wall turbulence. *J. Fluid Mech.*, 165: 163–199. , ,

86. Perry, A. E. and Marusic, I. 1995. A wall-wake model for the turbulence structure of boundary layers Part 1. Extension of the attached eddy hypothesis. *J. Fluid Mech.*, 298: 361–388. , ,

87. Rao, K. N., Narasimha, R. and Badri Narayanan, M. A. 1971. The

'bursting' phenomena in a turbulent boundary layer. *J. Fluid Mech.*, 48: 339–352.,

88. Robinson, S. K. 1991. Coherent motions in turbulent boundary layers. *Ann. Rev. Fluid Mech.*, 23: 601–639.,

89. Rodriguez, J. F. and Garcia, M. H. 2008. Laboratory measurements of 3-D flow patterns and turbulence in straight open channel with rough bed. *J. Hydraulic Res.*, 46(4): 454–465.,

90. Rogers, M. M. and Moin, P. 1987. The structure of the vorticity field in homogeneous turbulent flows. *J. Fluid Mech.*, 176: 33–66.,,

91. Roy, A. G., Langer, T. B., Lamarre, H. and Kirkbride, A. D. 2004. Size, shape and dynamics of large-scale turbulent flow structures in a gravel-bed river. *J. Fluid Mech.*, 500: 1–27.,

92. Sawford, B. 2001. Turbulent relative dispersion. *Ann. Rev. Fluid Mech.*, 33: 289–317.,,

93. Schoppa, W. and Hussain, F. 2002. Coherent structure generation in near-wall turbulence. *J. Fluid Mech.*, 453: 57–108.,,

94. Shvidchenko, A. B. and Pender, G. 2001. Large flow structures in a turbulent open channel flow. *J. Hydraulic Res.*, 39(1): 109–111.,

95. Smith, C. R. A synthesized model of the near-wall behavior in turbulent boundary layers. Proc. 8th Symp. Turbulence. Edited by: Zakin, J. and Patterson, G. pp.299–325. University of Missouri-Rolla.

96. Smits, A. J., McKeon, B. J. and Marusic, I. 2011. High Reynolds number wall turbulence. *Ann. Rev. Fluid Mech.*, 43: 353–375.,

97. Sukhodolov, A. N., Nikora, V. I. and Katolokov, V. M. 2011. Flow dynamics in alluvial channels: the legacy of Kirill V. Grishanin. *J. Hydraulic Res.*, 49: 285–292.,

98. Tamburrino, A. and Gulliver, J. S. 1999. Large flow structures in a turbulent open channel flow. *J. Hydraulic Res*, 37: 363–380.,,

99. Tamburrino, A. and Gulliver, J. S. 2007. Free-surface visualization of streamwise vortices in a channel flow. *Water Resour. Res.*, 43(W11410): 1–12.

100. Tardu, S. 1995. Characteristics of single and clusters of bursting events in the inner layer, Part 1: Vita events. *Exps Fluids*, 20: 112–124.,,

101. Theodorsen, T. Mechanism of turbulence. Proc. 2nd Midwest. Conf. Fluid Mechanics. March17–19. pp.1–19. Columbus: Ohio State University.

102. Toh, S. and Itano, T. 2005. Interaction between a large-scale structure and near-wall structures in channel flow. *J. Fluid Mech.*, 524: 249–262.,

103. Tomkins, C. D. and Adrian, R. J. 2005. Energetic spanwise modes in the logarithmic layer of a turbulent boundary layer. *J. Fluid Mech.*, 545: 141–162. ,

104. Townsend, A. A. 1976. *The structure of turbulent shear flow*, Cambridge UK: Cambridge University Press.

105. Velikaniv, M. A. 1958. *Channel processes (in Russian)*, Moscow: Fismatgiz.

106. Wallace, J. M., Eckelmann, H. and Brodkey, R. S. 1972. The wall region in turbulent shear flow. *J. Fluid Mech.*, 54: 39–48.

107. Wark, C. E. and Nagib, H. M. 1991. Experimental investigation of coherent structures in turbulent boundary layers. *J. Fluid Mech.*, 230: 183–208. ,

108. Willmarth, W. W. and Lu, S. S. 1972. Structure of the Reynolds stress near the wall. *J. Fluid Mech.*, 55: 65–92. ,

109. Yalin, M. S. 1992. *River mechanics*, Oxford UK: Pergamon Press.

110. Zagarola, M. V. and Smits, A. J. 1998. An-flow scaling of turbulent pipe flow. *J. Fluid Mech.*, 373: 33–79.

Chapter 7

HYDRAULIC STRUCTURES: A POSITIVE OUTLOOK INTO THE FUTURE

Willi H. Hager[1] and Robert M. Boes[2]

[1]IAHR Honorary Member), Professor, VAW, ETH Zurich, CH-8093 Zürich, Switzerland
[2]IAHR Member), Professor and Director, VAW, ETH Zurich, CH-8093 Zürich, Switzerland

ABSTRACT

Hydraulic structures are, and will remain, relevant in the future, given the enormous problems in water engineering to come. After having been a key topic up to the Second World War, they have lost attractiveness until recent years, given the advance of mainly river and environmental engineering. Following the recent developments in laboratory instrumentation, a number of problems in hydraulic structures can be solved mainly using experimentation, thereby accounting for the flow complexities given by the often spatial and highly turbulent flow structures, combined with multi-phase flow. An outlook into the future of hydraulic structures further reveals the importance of updated hydraulic book series, from which students and practicing engineers profit from the knowledge available among the International Association for Hydro-environment Engineering and Research Committees. As a conclusion, hydraulic structures will play a dominant role in the future, given the enormous but fascinating problems posed by the scarcity and abundance of water on our planet.

INTRODUCTION

Hydraulic structures may be defined as engineering elements used in water engineering allowing for improved or modified flow features as compared with natural water flow. They either improve the flow conveyance by reducing resistance, or are inserted in water bodies to deflect, expand, or contract a flow. These structures are therefore mainly made of massive concrete, given the often high forces to be expected due to increased velocities, the limited conveyance in an environment limited in space, and the high discharges to be expected mainly for design conditions. River engineering, in contrast, deals mainly with water-sediment flows, in which the flow boundaries are typically loose, and in which scour and deposition due to water flow are to be expected. To counter these deficiencies, concrete is also employed in river engineering, as at bridge piers or abutments, or rip rap is placed at vanes, or groins to improve river flow. The following does not deal with these elements, but only considers hydraulic structures in the proper sense.

Below, key historical developments are first summarized, the relevance of laboratory studies is highlighted, followed by an outline of current hydraulic problems and future research directions. The paper concludes with a discussion of the role played by the International Association for Hydro-environment Engineering and Research (IAHR) in the development of hydraulic structures as an important part of modern hydraulic research.

HYDRAULIC STRUCTURES: FROM THE NINETEENTH TO THE TWENTY-FIRST CENTURY

Up to about the end of the Second World War, research in hydraulics was mainly concerned with questions relating to hydraulic structures. It was realized in the nineteenth century that dams and reservoirs had important tasks in the society, relating to flood defence, energy production, and water supply. With the advance of dam engineering technologies, dams of taller and wider dimensions were built, thereby increasing both the discharge and the flow velocity (Schnitter 1994). Whereas fill dams around the turn of the twentieth century were mainly made of rockfill, earth-fill dams were erected following mainly the advents in geotechnical engineering and soil mechanics. The worldwide development of high dams above 100 m height after the Second World War is impressively illustrated by their total number increasing from then 30 to nearly 500 in 1990. Schnitter also states that over these decades, gravity and buttress dams were replaced by arch dams, and eventually by earth and rockfill dams. In terms of hydraulic structures, completely new problems arose, including cavitation damage, jet scour, or abrasion. These questions were at the time mainly studied by the countries with the leading dam activities, including the USA with their national authorities, the US Bureau of Reclamation, or the US Army Corps of Engineers. Further, the Soviets as the second world power also contributed to these issues by large national studies. During the 1960s and 1970s, new problems in hydraulic research were discussed, including density currents, withdrawal of water, stratification in large reservoirs, marine currents, and coastal engineering, so that hydraulic structures eventually lost their fresh touch in hydraulic engineering. With the advent of numerical computations in the 1970s and 1980s, questions relating to hydraulic structures were more and more neglected, because these problems were thought to be solved, and it appeared to be out of fashion to deal with flows having been studied by our grand-fathers. Another reason hardly stated was the complexity of flows in hydraulic structures, which remained unsolved with the then available techniques in computational fluid dynamics (CFD). A closer look reveals that these flows often are made up of at least two phases, they are highly turbulent, and often spatially three-dimensional (3D) and heterogeneous, so that the then available CFD techniques simply could not be applied. There were times in the 1990s, during which it was almost a shame to state "I am working in hydraulic structures," because colleagues thought that the only issues would be weir and gate flows, hydraulic jumps and energy dissipators, among others. It is true that these flows were almost fully understood in hydraulic practice, i.e. if the interest was solely in discharge or loss coefficients, submergence limits, or structural dimensions. However, it was overlooked that the overall hydraulic problems then are by far not solved.

During the past several years, hydraulic structures in the modern sense of the notion have experienced a revival. There are several causes for this positive development. Hydraulic instrumentation has greatly advanced, so that fundamental flow parameters can be better measured. Think, for instance, of air–water flows, for which probes are available to detect local air concentrations, bubble characteristics, mixture velocities, or boundary layer information. Think also of the advances of pressure cells by which the turbulent flow features are recorded. Think of laser-based techniques such as particle image velocimetry (PIV) and laser Doppler anemometry by which entire velocity fields are investigated with a non-intrusive approach, or of the Velocity Tracking methods to survey surface velocity fields. These techniques were developed during the past few decades, and in the meantime have reached a commercial availability, so that anybody interested in these processes can purchase standard instrumentation. As in many other fields of science, instrumentation has therefore had a dramatic advance in the recent past, so that parameters can be recorded that were hardly amenable only a few years ago, thereby influencing the current research directions significantly.

The second issue of relevance is the worldwide search for energy, stimulating the revival of hydraulic structures. Hydropower still is the most important renewable energy source, so that numerous countries exploit their potential. The People's Republic of China, for instance, has erected with the

Three Gorges Dam the largest hydropower development worldwide. India and Brazil also look confidently into the future in terms of hydropower, to satisfy their thirst for energy. A further motor of this energy was in addition the 2011 Fukushima Disaster, shedding a dramatic light on the future of nuclear energy. Some countries have announced their farewell from this energy source, which will have to be soon replaced with other energy systems. Among the various possibilities, including wave, wind, geothermal, or biomass energies, hydropower is known to be globally accessible, it is a well-developed technology, is often available and relatively safe as compared with others. Accordingly, there appears to be a revival of hydropower, and thus hydraulic structures.

A third reason for the revival of "hydraulic structures" is the need of experimental data in CFD. Going through the leading journals of hydraulic engineering reveals that many so-called hybrid studies are conducted to gather data for numerical calibration and validation. It can be stated that most advanced CFD codes need this input, to calibrate the many variables affecting the outcome. This is a highly positive development in terms of the profession, because a problem is no more attacked by either experimental *or* numerical approaches, but by a combined technique resulting in improved final conclusions of a certain problem. This trend was, for example, supported by the Journal of Hydraulic Engineering of the American Society of Civil Engineers, in which so-called Case Studies have attracted researchers in the past few years. These works are often in hydraulic structures, thereby stating their relevance in the modern hydraulics community.

A last reason to be detailed here is simply the immense importance of water on our planet. The statement "Without water life is impossible" may be considered basic, but it is true for a growing population, a society aiming at a certain living standard, and a mankind that deserves a proper life. Issues to be addressed include water supply, water treatment, water quality, adequate settings for the global water reserves, water availability during droughts, and water defence during floods. The latter are still an enormous challenge, and there is currently no safe method to counter the regularly returning misery due to water inequilibrium. In contrast to past centuries, the modern society requests water when needed, thereby imposing a strong requirement in safety. This is particularly related to hydraulic structures, given the large velocities and discharges involved, harnessing a valley or a shore to the point that the environment may be completely damaged, as experienced by the large dam disasters after the Second World War. The society thus supports our advances, but clearly demands that all hydraulic projects be safe in the sense that the potential damage remains under control. Accordingly, the requirements for hydraulic structures have dramatically increased, so that the major issues of all structures erected by engineers should be sufficiently checked. Relatively new researches as impulse waves have been established from this safety aspect. These waves, also referred to as mega-tsunamis, are generated by slides impacting large water bodies, as reservoirs. Due to the momentum exchange, waves higher than 100 m may be generated, with an enormous damage potential at run-up or run-over of opposite reservoir shores. The Fukushima Disaster has demonstrated the power of water, and the great human consequences following this incident.

As a consequence, hydraulic structures appear to be an important issue in the future of the hydraulic engineer, so that a revival of an old topic is likely to occur. It may be stated that the current knowledge seems to be sufficient to counter the negative effects of these man-made structures, and it can also be considered an invitation to all young members of the professional community: Join this field of work, you will find enormous challenges and exiting features! In passing, it should also be noted that the financing of hydraulic engineering in general, and hydraulic structures in particular, will likely increase in the future, given the enormous dependency on water, this important natural resource. These are good news for our profession, particularly for its younger representatives. Journals as the Journal of Hydraulic Research (JHR) will greatly add to this development, as does the IAHR, the only international association dealing with the timely issues on a global platform. In the following, the main future research directions are described, by which the "global water problem" appears to be partially solved, at least from the perspective of two hydraulic engineers, who work with pleasure and enthusiasm in this field. Before proceeding with the discussion of the current and forthcoming research problems, the role of hydraulic laboratories in hydraulic research will be stressed in the next section.

HYDRAULIC LABORATORY: NO TIME FOR RETIREMENT YET!

The first official hydraulic laboratory was founded in 1898 by Hubert Engels (1854–1945) at the University of Dresden, Germany. For the first time, he and his colleagues were able to see not only the flow close to the water surface as in a natural river, but also close to the channel bottom. Engels could easily study the effect of a certain element positioned into the flow, as a bridge pier, and at once realized the old fact that it fails not in, but against the flow direction, because of the horse-shoe scour features. His idea was quickly followed, so that three years later already four hydraulic laboratories existed in Germany. By 1929, John R. Freeman (1855–1932) edited a book devoted entirely to hydraulic laboratories, in which he presented examples from Europe, notably from Germany, then the leader. Given the enormous hydropower, water supply, and irrigation projects in the USA, numerous laboratories were erected there, so that after the Second World War, the Americans took over from Old Europe, and eventually became leaders in hydraulic engineering. During the 1960s and 1970s a hydraulics professor was hardly appointed in Europe if he had not been in the USA, then the Mecca of many scientific developments.

Despite these excellent advances, the hydraulic laboratories lost lots of potential with the advent of CFD. In the 1980s, engineers employing laboratory methods thought that they were out of fashion, given that the main hydraulic problems could now be solved with computers. That perception has been particularly influential at US universities and hydraulic establishments, so that many laboratories were either abandoned, or became in operational, along with the loss of its personnel. It may be stated that the USA and other countries have not recovered from that decision until now, and thus their output in terms of laboratory studies is limited. In contrast, other countries as Switzerland, for example, continued their laboratory activities along with the development of numerical studies, so that its research resources remain intact. To the astonishment of many colleagues, ETH Zurich has even inaugurated recently a new hydraulic laboratory, thereby stating the national interest in these works. The decision of ETH Zurich to support these activities in the twenty-first century clearly marks the interest in hydraulic engineering, an important aspect of the Alpine Countries. Similar facilities were recently installed, or are currently under construction, in Austria and Germany, adding thereby to the recent revival of hydraulic engineering. It will be interesting to see if other countries follow this trend.

It should be noted here that the VAW Hydraulic Laboratory does not only include the new building, but also the staff needed to run the facility. We are proud that ETH Zurich supports the laboratory strategy with seven permanent staff positions, including its laboratory head, the workshop head directing all the model works, two persons working mainly with PVC models, two persons working with model erection mainly, and one person responsible for all piping systems. When visiting other similar institutions, it is often this lack by which the laboratory appears incomplete. We have seen so many laboratories having their basic instrumentation, their channels, and their pumps, but laboratory personnel were missing. At first sight, this appears to be no problem, whereas the models standing in the laboratory made a different impression: They are either taken over from old material, or they are a poor representation of what would be expected. It appears obvious that a PhD student is simply unable to build up a proper hydraulic model, given his/her limited knowledge, timing, and absence of the required construction and manufacturing skills. Accordingly, many hydraulic studies that are still pursued typically require limited model work. However, lots of interesting and timely studies ask for a model designed and built by experts, namely the hydraulic laboratory workshop. This is identified as a reason of limited research activities in the field of what should be currently studied. Going through the research journals, there appear too many papers dealing with a problem that is not really relevant to the energy industry priorities, but simply was set up because it demanded limited model work, often done by the principal investigator himself or herself.

In summary, it should be stated that hydraulic laboratories constitute an inseparable element of hydraulic research, particularly its hydraulic structures branch. The future research activities should be governed by the hybrid approach, in which both experimental and numerical studies are dealt with, possibly even expanded with prototype observations. To achieve this goal, the hydraulic laboratory should provide a suit of modern facilities, including channels or piping systems, along with a workshop and laboratory personnel by which these facilities are built and maintained. Further, staff should be available specialized in the laboratory instrumentation, namely electronics personnel and mechanics. As stated, this may look superfluous at first sight, but the modern hydraulic laboratory is only operational and successful as a research platform with this standard.

MODERN HYDRAULIC PROBLEMS

Hydraulic structures often deal with open-channel flows. These can be either sub- or supercritical. Subcritical flows are often much simpler to visualize, given that they are normally one-phase flows. In contrast, supercritical flows often are two-phase air–water flows, complicating both the model set-up and flow observation. Below, both flow types are addressed, with an outlook to questions that appear to be important in the future.

Subcritical Flows

Given that flow depths in typical hydraulic models are below $h=0.30$ m, the typical flow velocity will be less than $(gh)^{1/2} = 1.7$ m/s, with g as gravity acceleration. Given these small flow velocities they offer no problems in terms of forces, or with instrumentation. It can be stated that one-dimensional (1D) flows can be considered solved, so that the research activities focus on the two-dimensional (2D) or even 3D aspects. In hydraulic engineering practice, the current approach mainly focuses on either 1D or 2D approximations, because a 3D solution is hardly considered an engineering approach. Therefore, let us concentrate on 2D hydraulic problems here.

Among the questions with subcritical flows, these in which the flow separates from the channel boundaries are of relevance, due to large zones of separation, in which turbulence effects become significant. Such a problem was recently studied at VAW, dealing with downstream fish migration along hydropower installations. To assure a safe fish passage across a low-head power plant, so-called louvres and bar racks were investigated, i.e. elements by which the fish is deviated from the turbine intake to a fish bypass to avoid turbine passage. A common hydraulic structure to achieve this goal involves a contraction element made up of a number of vanes, by which the fish is deflected to the lateral, thereby entering the fish bypass (Fig. 1). The flow encountered with these elements is clearly 3D, but it can be simplified to 2D, given the flow contraction up to the point of fish bypass, and the flow expansion further downstream. These elements involve a pro for fish and thus natural animal preservation, whereas it has a con for energy production, given the hydraulic losses experienced with louvre presence. In relation to this work, the associated hydraulic losses had to be established in terms of the many geometrical and hydraulic parameters. In addition, the flow structures were studied using PIV both in the horizontal and vertical directions, and by Acoustic Doppler Velocimetry (ADV). These data are used as the design basis for louvres and bar racks to be erected at run-of-river hydroelectric power plants, so that both environmentalists and hydraulic engineers are informed on their consequences. This problem falls into the category of environmental flows, but has essentially a hydraulic structures background.

Other problems to be solved with similar techniques include flow expansions, submerged weirs, or groins. In all cases, both ADV and PIV techniques are important and relatively straightforward to employ, given the almost horizontal water surface and the easy access from the channel side. Note that hydraulic structures are in most cases made up of rectangular channels, given their simple construction and the easy access. An important exception includes sewer mains, which are often made of circular conduits if their diameter is below, say 2 m. The access is then complicated, but the visualization problem is at least removed when using Plexiglas pipes. It can therefore be stated that subcritical flows in hydraulic structures commonly can be solved with the modern instrumentation, possibly accompanied by CFD studies to explore the turbulent flow structure.

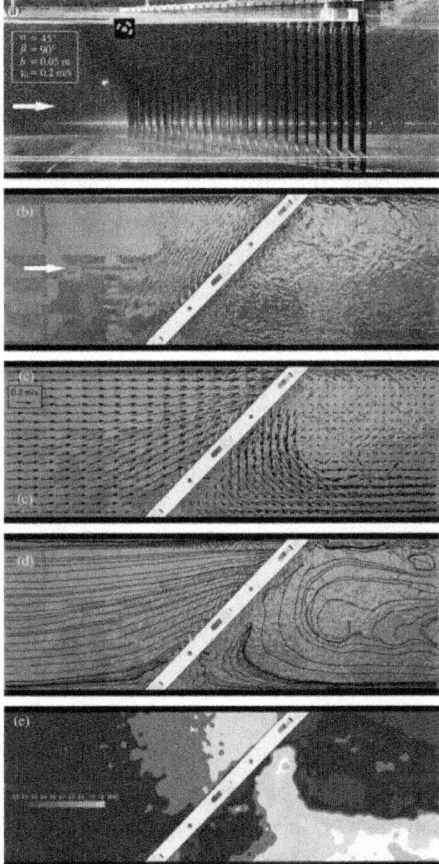

Figure 1: Louvre to deviate fish migrating from up- to downstream river reaches from turbines. Main screen angle 45°, slat angle 90°, slat distance 0.025 m, slat depth 0.05 m, rectangular slat thickness 0.005 m, approach flow velocity 0.20 m/s, approach flow depth 0.40 m, approach flow Froude number 0.10, approach flow Reynolds number 125,000 (a) side view, (b) 2D flow pattern, (c) plan vector field, (d) streamline map, (e) scalar map Source: Courtesy of VAW, ETH Zurich.

Supercritical Flows

As stated, these flows are often more challenging in laboratory studies than subcritical flows. In contemporary research journals, possibly only 10% of all papers dealing with hydraulic structures are devoted to supercritical flows, whereas the majority of studies refers to subcritical flow. This issue may be explained by various factors mainly relating to: (1) flow generation, (2) flow control, (3) model set-up, (4) data acquisition, and (5) scale effects. Below, the relevance of supercritical flows in hydraulic structures is highlighted first, followed by implications of the above factors.

In nature, except for extremely small flow depths as under rainfall, supercritical flow rarely occurs. These are therefore directed to man-made structures, as chutes, tunnels, or intakes. Given the large roughness of Mountain Rivers, even there the flow is normally subcritical, or transcritical at most over short flow reaches. Supercritical flows are mainly observed in man-made structures, as in chutes, tunnels, or intakes, generated under large heads, as typically below dams. Sediment effects normally are of no concern, except, e.g. at bypass tunnels of reservoirs, to avoid reservoir sedimentation.

Supercritical flow in hydraulic structures therefore commonly is a two-phase air–water flow. The air phase significantly complicates the study of the flow structure, given difficulties with the optical access and the large noise development. The latter became a nuisance, for instance, while investigating plunge pool scour, even at laboratory scale. In general, the noise of water flow can be related to its air content, as is experienced easily from flat-land and Mountain Rivers. The second challenge of these flows are so-called shocks or shock waves; these are generated whenever any disturbance is applied to the flow, including local changes of the bottom profile (as drops or crests), of the cross-section (as expansions or contractions), of the roughness structure (as transitions from rough to smooth, or vice-versa), or of lateral discharge additions or reductions (as for side channels or side weirs). Moreover, shocks also appear under sudden temporal changes, mainly by the addition or reduction of discharge. Shocks have received, at least in hydraulic structures, only a limited interest, the reasons being stated below. They are normally standing gravity waves causing local water surface extrema by which a chute has to be adjusted in height, or a tunnel be of larger diameter, so that free surface flow is conserved along the entire tunnel. If the flow is not limited to the free surface capacity, then overtopping occurs in chutes, or conduits choke, thereby generating a sudden transition from free surface to two-phase pressurized flow. The latter has to be strictly prevented, because extreme pressure fluctuations or air pocket flow may seriously damage a conduit. Accordingly, the effects of choking flow need to be explored in more detail.

As to the generation of supercritical flow, few advances have been made. With a gate or a sloping channel onto a horizontal tail water portion, large shocks result because of gate, or bottom deflection presence, perturbing massively the supercritical flow structure. Therefore, VAW, among others, developed the so-called jet-box more than 20 years ago, by which an almost perfect transition between the pressurized approach flow and the free surface tail water flow results. The jet-box is based on the principle of flow straighteners thereby producing a rectangular (or even circular) jet without any notable flow perturbations. This element is largely employed at VAW whenever supercritical flow studies are conducted.

The second issue relates to the flow control. A channel in which supercritical flow is examined should have an almost perfect shape, without any joints protruding into the flow, or width variations creating undesirable shocks. The quality requirements of these chutes therefore are high in terms of fabrication and maintenance, thereby asking for an excellent laboratory staff.

The third issue involves the model set-up. Recently, VAW conducted a study on the effect of triangular pyramids deflecting a high-speed jet from the approach flow direction. Both the approach flow and the elements inserted in the channel need to be of high finish, because the jet features cannot be studied otherwise. Consider also the additional challenges due to spray formation and jet impact onto the tail water, further complicating the experimentation. Note, therefore, that perturbations due to model inaccuracies and the approach flow should be much below these due to shocks, because the shock effect will be smeared over with the former perturbations, resulting in poor data.

The data acquisition of supercritical flows is often problematic. Given the high velocities, standard instrumentation cannot be used in hydraulic laboratories due to the high forces on these instruments. Consider, for instance, flow over stepped spillways at flow depths of 0.1–0.2 m, and velocities of say 5 m/s. The force onto a 10 mm × 10 mm probe then amounts to 4.4 N. Therefore, VAW developed decades ago special probe carriages allowing for work under these circumstances. Today, all these instruments are connected with Lab View, by which the required test procedure can be programmed. A further complication experienced during the study of these processes is the extremely rough flow surface, including bubbles and air pockets, hindering the use of PIV or ADV. Even propellers to measure a certain point velocity come to their limits, so that standard measurement often relates only to depth readings via a point gage. This information is sufficient in usual cases, because hydraulic engineers are mainly interested in the main standing wave patterns, and not so much into the velocity field. Pressure readings along the flow boundaries also come to the limit, given the air–water mixture flow and problems with accurate pressure readings due to the turbulent flow pattern. Fibre-optical probes enable to measure local void fractions, bubble size, and mixture velocities even in high-speed air–water flows of hydraulic models (Boes and Hager 2003). The optical access using CCD cameras was recently employed to estimate local air concentrations in a two-phase flow, but this field is open for improved techniques. Last but not least, scale effects need to be carefully assessed in supercritical flows. Given the limited pump capacities, too narrow channels, or small flow depths may result in incorrect data governed by mainly viscosity and surface tension effects (Fig. 2). Heller (2011) gives an impressive outline on these aspects, stating limit values for common hydraulic problems in terms of minimum flow depths to be applied. His work describes the flow of water and air, which is subjected to modified limitations if other fluids were used. In general, the VAW experience suggests that to exclude major scale effects in open-channel flows, the minimum flow depths should be around 50 mm. Minimum channel widths are of the order of 0.30 m, because the 3D effect is otherwise lost. Regarding surface tension and viscosity effects in stepped spillway flow, Boes and Hager (2003) propose limiting values of both the Weber and Reynolds numbers. Supercritical flow studies thus require a minimum of hydraulic limitations to produce reliable data.

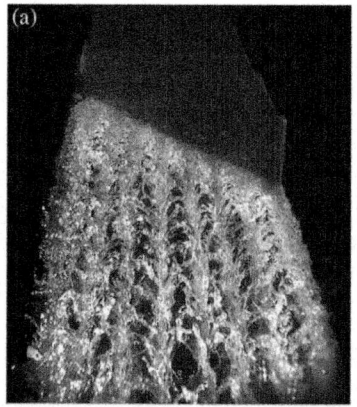

Figure 2: Scale effects on Kárahnjúkar chute in Iceland (a) model observation with 1:45 scale, (b) prototype flow (Pfister*et al.* 2011)

PERTINENT PROBLEMS

In the following, problems of interest in relation to hydraulic structures are described, based mainly on the Authors' experience. These can be summarized as two-phase air–water flows, vorticity flows, jet flows, and transitional free surface to pressurized flows.

Air–Water High-Speed Flows

Consider Fig. 3, showing flow across a bottom outlet. This hydraulic structure involves a tunnel connecting a reservoir with the tail water via a high-pressure gate transforming pressurized flow into free surface flow. This structure represents an important safety element allowing at the limit for the complete reservoir emptying. It is also often used for removal of the reservoir sedimentation. For a reservoir of typical height of 100 m, the outlet velocity amounts to 45 m/s, or 160 km/h. Bottom outlets thus produce the highest velocities in hydraulic engineering, yet they have hardly been investigated. From over the past 50 years, less than 10 studies are available, pointing at an important gap in the hydraulic research agenda. It should include studies of gate outflow, flow aeration to avoid choking flow, cavitation damage along the tunnel, and effects of high-speed tunnel flow in terms of discharge capacity.

Figure 3: High-speed flow at bottom outlet (Vischer and Hager 1998)

A second problem of relevance in this category relates to chutes. Given the high velocities as in bottom outlets, the analogous concerns apply here, namely cavitation damage and flow bulking (Falvey 1980, 1990, Chanson 1989, 1993, Wood 1991, Boes 2012, Frizell *et al.* 2013). It is currently state-of-the-art to employ aerators to counter cavitation damage, but the aeration features of typical chute aerators were only recently established (Pfister and Hager 2010, Pfister *et al.* 2011). It was found that standard chute aerators have a limited field of influence, and that a second or even third aerator must be provided for long chutes in terms of the de-aeration characteristics and typical flow depth. As shown in Fig. 4, depicting air concentration contours, the air–water mixture flow downstream of the aerator massively detrains air upon jet impact onto the chute bottom. Therefore, a further aerator must be provided to protect the chute from cavitation damage as soon as the air bottom concentration falls below the minimum value of the element considered. Adding more air at the first aerator does not solve the problem, because a certain flow has a maximum air transport capacity, and only flow bulking would result.\

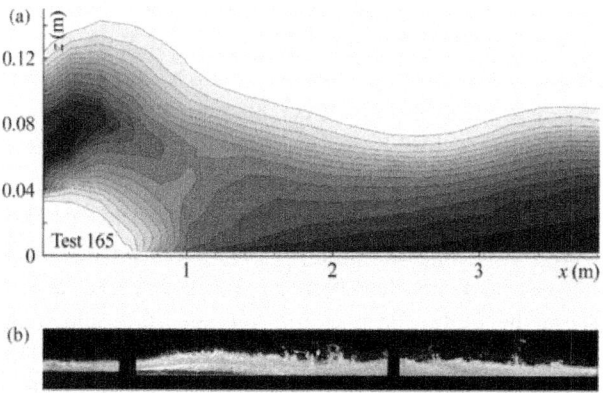

Figure 4: (a) Air concentration contours, (b) flow features in tail water of chute aerator (Pfister *et al*. 2011)

A third issue that remains poorly investigated is ski jump flow, deflecting high-speed chute flow away from a dam into the tail water. Because stilling basins are limited in terms of approach flow velocities to some 20 m/s, ski jumps were introduced before the Second World War, yet most studies were directed to particular project studies (Khatsuria 2004). A decade ago, VAW launched a research series focused on this issue from a general perspective. A number of studies (Heller *et al*. 2005, Schmocker *et al*. 2008, Steiner *et al*. 2008, Pfister *et al*. 2014) aimed at developing expressions for the jet trajectories, the optimum ski jump geometry, the effects of the relative deflection radius and the approach flow Froude number, among others, on the bucket pressure distribution, the jet geometry, and the hydraulic impact features. Currently the main problems are solved, but there remain additional questions to be considered in the future, relating, e.g. to elements positioned on the bucket, as proposed by Chinese engineers (Hager and Schleiss 2009), to further diffuse the jet flow, and to reduce the hydraulic impact onto the plunge pool in the tailwater with jet expanding elements. In addition, a number of problems have also been identified in the plunge pool, namely effects of the air–water approach flow on the scour, 3D scour issues, spray development and its reduction, and the effect of a rocky matrix as compared with granular sediment commonly used in hydraulic tests. Figure 5 highlights flows associated with a ski jump.

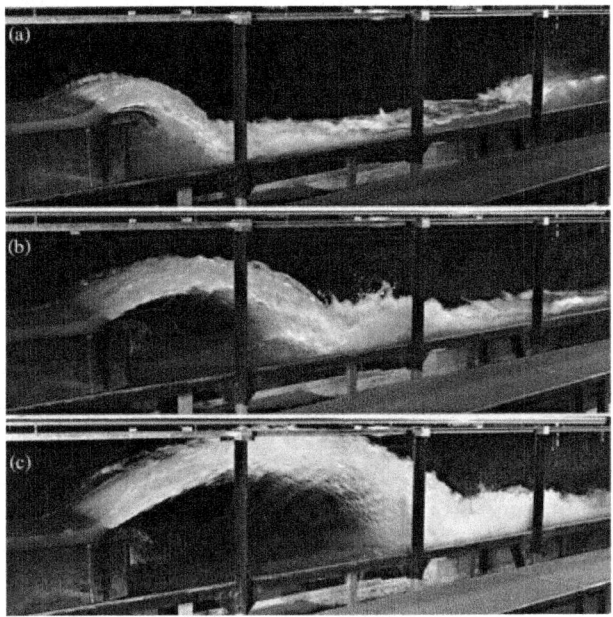

Figure 5: Ski jump as efficient energy dissipator under high-speed approach flow, approach flow Froude number F_o = (a) 3, (b) 5, (c) 7 (Steiner *et al.* 2008).

Free Surface to Pressurized Flows

Two flows are discussed under this category, namely intakes and manhole flows. Intakes to turbines for hydropower production are characterized by a temporally varied approach flow elevation. Typically, the reservoir level is subject to changes with the seasonal discharge regime. If the reservoir submergence above the intake is sufficiently high, then no concerns occur; however, at a certain lower reservoir level, vortices above the intake become so strong that incipient air entrainment into the intake results. If the reservoir level is even lower than the threshold value, air is entrained from the reservoir surface, by which poor intake conditions result (Jiming *et al.* 2000, Yıldırım *et al.* 2011, Mortensen *et al.* 2012, Suerich-Gulick *et al.* 2014). These include a significant reduction of turbine efficiency, given that already a few per cents of air discharge compared with the water discharge yield a drastic reduction of turbine performance. Further, the air flow into the intake tunnel may cause pressure fluctuations, air accumulations, or even air backflow, which has to be avoided under all circumstances (Gulliver *et al.* 1986, Gulliver and Rindels

1987, Ma *et al.* 2000, Liu and Yang 2013). To counter these effects, relatively few advances were made in the past. The design recommendations of Gordon (1970) and a manual by Knauss (1987) are available, but only few details on air–water flow at intakes are currently available (Padmanabhan 1984, Möller *et al.*2014).

To improve the current design guidelines, VAW conducted studies, among which is that of Glauser and Wickenhäuser (2009) dealing with air venting of tunnels, and that of Möller *et al.* (2014), in which air entrainment is studied. The latter intake involves a sharp-crested pipe protruding into a rectangular container much larger than the pipe diameter. Figure 6a shows the model set-up, including the container, the Plexiglas pipe of diameter D=0.40 m, and the intake vortex with a small vortex core entraining air into the pipe that is partially accumulated at the inlet, but then transported downstream, as previously described. The air discharge was measured with a venting pipe mounted at the tailwater (Fig. 6b), resulting in an expression between the ratio of the air Q_a to the water Q_w intake discharges versus the relative intake flow depth h/D and the pipe Froude number $F_D = V/(gD)^{1/2}$, with V as average cross-sectional pipe flow velocity.

Figure 6: Air discharge Q_a at h/D=1.5, F_D=0.8 (a) air-entraining vortex, (b) de-aeration at tailwater pipe (Möller *et al.* 2014)

Another typical problem of free surface to pressurized flows occurs at manholes of sewer mains at large discharge. Figure 7shows a 45° bend manhole under supercritical approach flow, by which transition to pressurized manhole flow results if the discharge is in excess of the manhole capacity discharge. Typically, the axial bend radius is $3D$, with D as the sewer diameter, resulting in a strong flow deflection, as is seen in Fig. 7a, with the forward flow attaching to the outer bend portion, and a wide recirculating flow along the inner bend portion, by which the bend flow becomes hydraulically poor. Note from Fig. 7b the excessive height of the standing wave along the outer bend portion impacting the manhole end wall, causing the breakdown of the supercritical flow across the manhole.

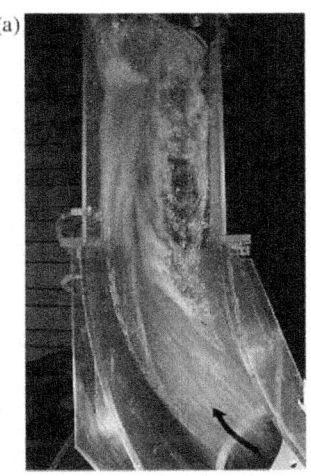

(b)

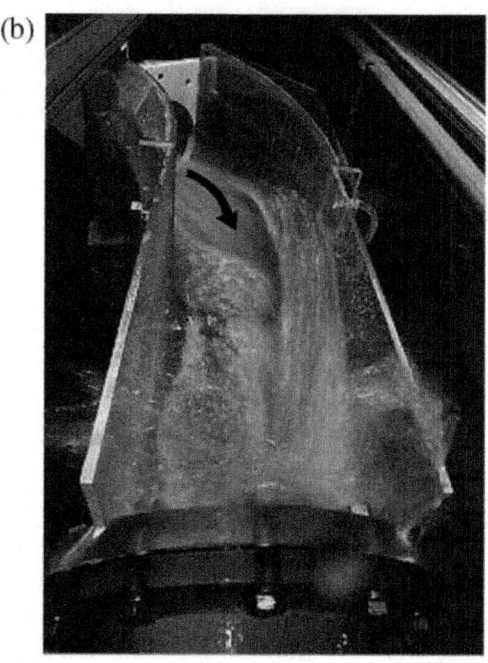

Figure 7: Flow in 45° bend manhole (a) plan view, (b) view from downstream.

This problem was solved using hydraulic laboratory experimentation, given the 3D flow structure, the non-hydrostatic pressure distribution, and the two-phase air–water flow. It was found that, in contrast to current design, the outer manhole wall should act as flow boundary, thereby not allowing for water to flow onto the manhole benches. Further, again in contrast to standard bend manhole design, the manhole should contain a $2D$ long expansion downstream of flow deflection so that the maximum shock wave is reduced in height up to the manhole exit. Note from Fig. 8a the improved manhole design, with the manhole access confined entirely along the inner bend portion, and the manhole extension to increase the manhole discharge capacity. Figure 8b shows a streamwise section across the bend manhole, with h_o as the approach flow depth, and V_o as the average cross-sectional approach flow velocity, so that the approach flow Froude number in the U-shaped manhole channel is $F_o = Q/(g D^2 h_o^3)^{1/2}$. Semi-empirical expressions for the maximum (subscript M) and minimum (subscript m) flow depths h_M and h_m, and the corresponding angle θ_M in terms of the approach flow filling ratio $y_o = h_o/D$ and F_o were developed. Further, the swell height h_s was determined, given its relevance for entirely supercritical flow across the hydraulic structure. Note that manhole choking occurs if $h_s > D$, because of flow impact onto the manhole outlet wall.

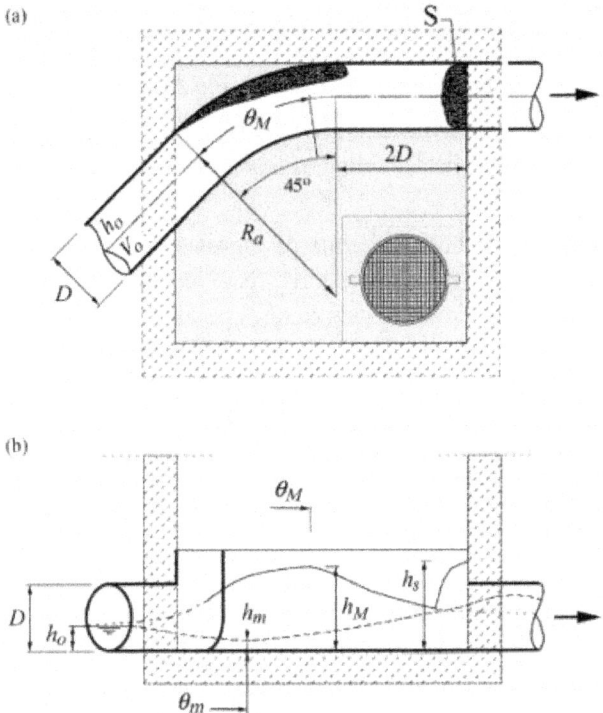

Figure 8: Bend manhole with manhole extension of length $2D$ (a) plan with S as swell, (b) section with flow profiles along (—) outer wall, (- - -) inner wall (Hager 2010).

Flow choking must be avoided, so that supercritical approach flow breaks down neither along the manhole nor in the tail water. Experimentation indicated that two conditions must be satisfied: (1) approach flow filling ratio $y_o < 2/3$, and (2) manhole capacity Froude number $Fc = Q/(gD^5)^{1/2} < 0.90$. The first condition imposes a significant reduction of the standard sewer filling ratio of typically 85%, whereas the second condition implies a relation between the discharge and the sewer diameter. Note that no other conditions apply if the design shown in Fig. 8 is employed, independent of the bend deflection angle. This guideline allows for a straightforward and simple manhole design.

OUTLOOK INTO THE FUTURE

As stated, the expected research activities in hydraulic structures are diverse and challenging. Below, a list of topics is proposed to highlight the issues requiring fundamental study, addressing current and emergent problems (without any ranking in priority). These topics have to be considered in a generalized way first, and then detailed in relation to a specific engineering project. Both physical and computational modelling, along with field studies will be of importance in the future. This compound approach should result in hydraulically correct and economically feasible solutions.

Based on past experience with hydraulic structures and the expected pertinent problems, the list of relevant research topics in hydraulic structures includes:

- *Bottom outlets* both in terms of high-speed flow and particularly air flow features. These flows have received insufficient attention in the past due to experimental challenges posed by the instrumentation, and poor access to tunnel flows. Notable studies in this field were conducted by Wagner (1967), Sharma (1976), Blind (1985), Naudascher and Rockwell (1994), Sagar (1995), and Speerli and Hager (2000). They underpin urgent actions required to further investigate these important safety outlets, including issues associated with reservoir sedimentation.

- *Sediment bypass tunnels* as applied at dams with a heavy sediment flux including large boulders combined with high-speed flow, by which a large portion of mainly bed-load is bypassed, so that reservoir sedimentation is reduced and the tailwater river supplied with sediment, thereby re-establishing sediment continuity. This concept was rarely applied so far, except for mainly Japan and the Alpine countries, due to the investment and substantial maintenance cost resulting from significant hydro-abrasion of the tunnel invert. Few studies have been undertaken by which the abrasion of the tunnel bottom is sustainably controlled (Jacobs *et al.* 2001, Sumi *et al.* 2004, Sumi and Kantoush 2011, Auel and Boes 2012, Auel *et al.* 2014).

- *Desilting facilities* by which sediment upstream of turbines and other hydraulic machinery is separated from water flow. Given the enormous abrasion potential of sediment particles down to sizes as small as silt, the performance and longevity of hydraulic machinery is significantly increased if essentially a pure water flow instead of sediment-laden flow results from efficient and economically well-designed desilting facilities (Garde *et al.*1990, Olsen and Skoglund 1994, Cheng 1997, Ranga Raju *et al.* 1999, Padhy and Saini 2009).

- *Roughness problem* in hydraulic engineering, typically along chutes and tunnels, by which the flow undergoes variations that have to be understood from the hydraulic perspective, to avoid choking flow, sediment deposition, and low-pressure zones. Most of the current hydraulic models involve smooth boundary surfaces, while the effect of increased roughness by systematic roughness variation has been hardly considered in the past (Schröder 1990).

- *Tunnel intakes* as a spatial transition from a reservoir or a river to a tunnel, forcing free surface tunnel flow to avoid cavity or pressurized air–water flow regions. Given that tunnel flow is normally supercritical, shock waves are to be expected at the tunnel intake as well as further downstream, e.g. at tunnel bends. These shocks tend to generate a transition from free surface to pressurized two-phase flow, invalidating the concept of free surface tunnel flow (Vischer and Hager 1998).

- *Drop shafts* into pressurized hydraulic systems, by which again the flow undergoes perturbations by air presence. The air from the system can only be evacuated by de-aeration devices, for which the design information is currently poor. Hydraulic experimentation must be conducted at sufficiently large models to reduce scale effects, so that the laboratory results are correctly up-scaled (Novak and Cabelka 1981, Ervine and Himmo 1984).

- *Vortices at tunnel intakes*, by which the tunnel flow is composed of an air–water discharge, with the problems addressed previously in terms of tunnel flow stability, air evacuation, and vibrations. Given the normally small amount of air, challenges in instrumentation and model set-up are posed. So far, extremely few advances of this problem in relation to hydraulic structures have been made (Anwar 1966, Gordon 1970, Anwar *et al.* 1978, Gulliver *et al.* 1986, Gulliver and Rindels 1987, Knauss 1987).

- High-speed outlets typically downstream of bottom outlets, involving so-called flip buckets by which the flow is lifted into the atmosphere and simultaneously deflected in the plan. Given the enormous flow velocity, the common ski jump does not apply. A successful flipping action is based on additional elements preventing cavitation and other damaging problems. So far, this research topic has hardly been considered (Mason 1989, Lucas *et al.* 2013).

- Submerged flow downstream of weir and gate structures at low-head hydropower facilities. As compared with the free flow mode, the complexities of submerged flows are high, resulting in flow separation, vibration, low-pressure flow zones, and complicated spatial flow features, along with air entrainment along shear zones. Given these complexities, and the *in situ* particularities, few researches have been conducted in this field (Clemmens *et al.* 2003, Castro-Orgaz *et al.* 2013).

- Fish migration facilities at low-head power plants both in the up- and particularly in the downstream direction. Apart from the environmental problems, purely hydraulic features have to be considered in the selection of the appropriate means to facilitate the safe passage of fish across turbines, thereby also accounting for the adequate reduction of power output from the hydropower installation. Facilities involving louvres and other suitable rack elements, by which fish is bypassed from turbines, are considered the only means to return fish safely to rivers, from where they were partly banned once hydropower infrastructure was installed (Kynard and Horgan 2001, Raynal *et al.* 2013).

The IAHR *JHR* would be the most appropriate place for the dissemination of fundamental studies underpinning the above research topics, chiefly those focusing on the theoretical issues and innovative modelling approaches. As to the broadcasting of the applied work and project-specific findings, authors could consider the IAHR *Journal of Applied Water Engineering and Research* (*JAWER*) that has been specifically established to serve this purpose.

IAHR AND HYDRAULIC STRUCTURES

IAHR was originally named the International Association for Hydraulic Structures Research, and the shorter form without "Structures" was only adopted in the early 1950s. This fact sheds again light on the relevance of hydraulic structures during at least the initial IAHR era. In the 1980s, the late Prof. Eduard Naudascher, Karlsruhe Technical University, developed the IAHR Hydraulic Structures Design Manual series, in which a total of nine volumes until 1995 were published, but from then no more additions followed. This is a pity because continuation of this initiative would remove a currently often stated IAHR deficiency: the strength of IAHR connection with hydraulic practice remains to be improved. It should also be stated that, following the 2009 IAHR name change, it was intended to present either a monograph, or organize a Workshop by each IAHR Committee. Regrettably, this has not been realized in most cases so far. We think that these activities would greatly improve the stated IAHR deficiency, particularly if timely topics as River Engineering, Hydraulic Structures, or Environmental Flows would become book projects within the IAHR Committees. These topics include indeed not only a research attraction, but also often constitute real engineering problems. Accordingly, these activities rank in the vision not only of IAHR, but also in its Hydraulic Structures Committee. It is evident that the authorship of books has greatly lost attractiveness, mainly for younger researchers, because these are interested in peer-reviewed journal papers, for which they are given adequate credits in terms of the impact factor. This is certainly a good development on the one hand, but has massive disadvantages on the other hand, which are felt particularly by the practicing hydraulic engineer. He or she is hardly ready to read papers because their conclusions often relate to a particular procedure, or a minor modification of a known fact. In contrast, their work would greatly profit from updated monographs, in which the current knowledge is reviewed, and in which examples are presented allowing for the solution of daily hydraulic problems. A search on the currently available books in hydraulic structures, and hydraulics, reveals that only few have been published within the past 10 years (e.g. Chanson 2004, Khatsuria 2004, Novak *et al.* 2006, 2010, Hager and Schleiss 2009). Given the large projects in water engineering to be expected in the near future, as mentioned above, this is indeed astonishing. The only answer can be that the current hydraulic researcher (who often is IAHR Member) does not want to take the time for this urgent task. We think here particularly of researchers in excess of say 40–45 years of age, whose careers have settled in a top position within an academic environment. They appear to be unwilling or unable due to diverse reasons to devote a part of their time to this urgent task to support the hydraulic community, and to make available their knowledge to younger members and to the hydraulic practice. It appears that there exists currently no simple "recipe" to initiate these works...

Looking through the *JHR*, the IAHR flagship journal, it also becomes evident that "hydraulic structures" have a great impact there. In 2012, for example, a total of 64 *JHR* papers were published, of which more than a third deal with the present topic. Given the basic requirements of novelty, originality, and future orientation, most of these papers have, as stated, "only" a limited impact on the broad hydraulic community. We would like to stress that these statements are made in relation to the future hydraulic community, which does not only include scientists, but also students and practitioners, who could more profit from IAHR if this attitude were modified. Steps should therefore be taken by the IAHR Council if the dream of a wide hydraulics community should come true, so that a modern IAHR book series would certainly add to this aspect.

CONCLUSIONS

Hydraulic structures are one of the many topics dealt with in hydraulic engineering. Given the current global challenges including water scarcity and abundance in the form of floods, hydraulic structures will have a revival to satisfy the needs of the planet Earth, of which water is a fundamental element. It is demonstrated herein that hydraulic structures for a time have lost attractiveness among the hydraulics community for one or the other reasons, but a revival will be evident in the near future. Accordingly, it is important to prepare for this change not only by conducting research, but also in developing the knowledge for all individuals involved, namely students, the practicing water engineers, and the water managers.

The knowledge in hydraulic structures is well advanced mainly in subcritical flows, but is still limited in supercritical flows and transitions from free surface to pressurized flows. Typical developments in these three categories are outlined, with both numerical and mainly experimental approaches described. The latter two topics offer young students a great future "playground", in which novel ideas will be necessary to solve the most relevant problems. The Authors' vision relating to hydraulic structures is thus the hybrid approach using all available means, including numerical, experimental, and prototype studies, thereby highlighting the need for modern hydraulic laboratories. Further, it is believed, that peer-review journal papers should be complimented with updated design manuals as a service to the hydraulics community, by which both the engineers as also IAHR will profit.

REFERENCES

1. Anwar, H.O. (1966). Formation of a weak vortex. *J. Hydraulic Res.* 4(1), 1–16. doi: 10.1080/00221686609500089

2. Anwar, H.O., Weller, J.A., Amphlett, M.B. (1978). Similarity of free vortex at horizontal intake. *J. Hydraulic Res.* 16(2), 95–105. doi: 10.1080/00221687809499623 ,

3. Auel, C., Boes, R.M. (2012). Sustainable sediment management using sediment bypass tunnels. Proc. 24th *ICOLD Congress* Kyoto Q92 (R16), 224–241.

4. Auel, C., Albayrak, I., Boes, R.M. (2014). Turbulence characteristics in supercritical open-channel flows: Effects of Froude number and aspect ratio. *J. Hydraulic Eng.* 140(4), 04014004. doi: 10.1061/(ASCE) HY.1943-7900.0000841 ,

5. Blind, H. (1985). Design criteria for reservoir bottom outlets. *Water Power Dam Constr.* 37(7), 30–33.

6. Boes, R.M. (2012). Guidelines on the design and hydraulic characteristics of stepped spillways. Proc. 24th *ICOLD Congress* Kyoto Q94(R15), 203–220.

7. Boes, R.M., Hager, W.H. (2003). Two-phase flow characteristics of stepped spillways. *J. Hydraulic Eng.* 129(9), 661–670. doi:10.1061/ (ASCE)0733-9429(2003)129:9(661)

8. Castro-Orgaz, O., Mateos, L., Dey, S. (2013). Revisiting the energy-momentum method for rating vertical sluice gates under submerged flow conditions.. *J. Irrig. Drain. Eng.* 139(4), 325–335. doi: 10.1061/(ASCE) IR.1943-4774.0000552 ,

9. Chanson, H. (1989). Flow downstream of an aerator: Aerator spacing. *J. Hydraulic Res.* 27(4), 519–536. doi:10.1080/00221688909499127 , ,

10. Chanson, H. (1993). Self-aerated flows on chutes and spillways. *J. Hydraulic Eng.* 119(2), 220–243. doi: 10.1061/(ASCE)0733-9429(1993)119:2(220) ,

11. Chanson, H. (2004). *The hydraulics of open channel flows: An introduction.* Elsevier-Butterworth-Heinemann, Oxford, UK.

12. Cheng, N.-S. (1997). Simplified settling velocity formula for sediment particles. *J. Hydraulic Eng.* 123(2), 149–152. doi:10.1061/(ASCE)0733-9429(1997)123:2(149) ,

13. Clemmens, A.J., Strelkoff, T.S., Replogle, J.A. (2003). Calibration of submerged radial gates. *J. Hydraulic Eng.* 129(9), 680–687. doi: 10.1061/ (ASCE)0733-9429(2003)129:9(680) ,

14. Ervine, D.A., Himmo, S.K. (1984). Modelling the behaviour of air pockets in closed conduit hydraulic systems. *Scale. Eff. Model. Hydraulic Struct.* 4(15), 1–12.

15. Falvey, H.T. (1980). *Air-water flows in hydraulic systems.* Engineering Monograph 41. US Bureau of Reclamation, Denver, CO.

16. Falvey, H.T. (1990). *Cavitation in chutes and spillways.* Engineering Monograph 42. US Bureau of Reclamation, Denver, CO.

17. Frizell, K.W., Renna, F.M., Matos, J. (2013). Cavitation potential of flow on stepped spillways. *J. Hydraulic Eng.* 139(6), 630–636. doi: 10.1061/ (ASCE)HY.1943-7900.0000715 ,

18. Garde, R.J., Ranga Raju, K.G., Sujudi, A.W.R. (1990). Design of settling basins. *J. Hydraulic Res.* 28(1), 81–91. doi:10.1080/00221689009499148 , ,

19. Glauser, S., Wickenhäuser, M. (2009). Bubble movement in downward-inclined pipes. *J. Hydraulic Eng.* 135(11), 1012–1015. doi: 10.1061/ (ASCE)HY.1943-7900.0000093 ,

20. Gordon, J.L. (1970). Vortices at intakes. *Water Power* 22(4), 137–138.

21. Gulliver, J.S., Rindels, A.J. (1987). Weak vortices at vertical intakes. *J. Hydraulic Eng.* 113(9), 1101–1116. doi:10.1061/(ASCE)0733-9429(1987)113:9(1101)

22. Gulliver, J.S., Rindels, A.J., Lindblom, K.C. (1986). Designing intakes

to avoid free-surface vortices. *Water Power Dam Constr.*38(9), 24–28.

23. Hager, W.H. (2010). *Wastewater hydraulics: Theory and practice.* Springer, Berlin.

24. Hager, W.H., Schleiss, A.J. (2009). *Constructions hydrauliques: Ecoulements stationnaires* (Hydraulic structures: Steady flows). PPUR, Lausanne in French.

25. Heller, V. (2011). Scale effects in physical hydraulic engineering models. *J. Hydraulic Res.* 49(3), 293–306. doi:10.1080/00221686.2011.578914 ,

26. Heller, V., Hager, W.H., Minor, H.-E. (2005). Ski jump hydraulics. *J. Hydraulic Eng.* 131(5), 347–355. doi: 10.1061/(ASCE)0733-9429(2005)131:5(347) ,

27. Jacobs, F., Winkler, K., Hunkeler, F., Volkart, P. (2001). Betonabrasion im Wasserbau (Concrete abrasion at hydraulic structures). In *VAW Mitteilung* 168. H.-E. Minor, ed. VAW, ETH Zurich, Zürich, Switzerland.

28. Jiming, M., Yuanbo, L., Jitang, H. (2000). Minimum submergence before double-entrance pressure intakes. *J. Hydraulic Eng.*126 (8), 628–631. doi: 10.1061/(ASCE)0733-9429(2000)126:8(628)

29. Khatsuria, R.M. (2004). *Hydraulics of spillways and energy dissipators.* Taylor & Francis, Chichester.

30. Knauss, J. (1987). Swirling flow problems at intakes. *IAHR Hydraulic Structures Design Manual* 1. Balkema, Rotterdam.

31. Kynard, B., Horgan, M. (2001). Guidance of Yearling Shortnose and Pallid Sturgeon using vertical bar rack and louver arrays. *N. Am. J. Fisheries Manag.* 21(3), 561–570. doi: 10.1577/1548-8675(2001)021<0561:GOY SAP>2.0.CO;2 , ,

32. Liu, T., Yang, J. (2013). Experimental studies of air pocket movement in a pressurized spillway conduit. *J. Hydraulic Res.*51 (3), 265–272. doi: 10.1080/00221686.2013.777371 ,

33. Lucas, J., Hager, W.H., Boes, R.M. (2013). Deflector effect on chute flow. *J. Hydraulic Eng.* 139(4), 444–449. doi:10.1061/(ASCE)HY.1943-7900.0000652 ,

34. Ma, J.M., Liang, Y.B., Huang, J.T. (2000). Minimum submergence before double-entrance pressure intakes. *J. Hydraulic Eng.*126 (8), 628–631. doi: 10.1061/(ASCE)0733-9429(2000)126:8(628)

35. Mason, P.J. (1989). Practical guidelines for the design of flip buckets and plunge pools. *Water Power Dam Constr.* 45(9/10), 40–45.

36. Möller, G., Detert, M., Boes, R.M. (2014). Vortex-induced air entrainment rates at intakes. *J. Hydraulic Eng.* 140 (submitted).

37. Mortensen, J.D., Barfuss, S.L., Tullis, B.P. (2012). Effects of hydraulic jump location on air entrainment in closed conduits. *J. Hydraulic Res.* 50(3), 298–303. doi: 10.1080/00221686.2012.670008 ,

38. Naudascher, E., Rockwell, D. (1994). Flow induced vibrations. In *IAHR Hydraulic Structures Design Manual* 7. Balkema, Rotterdam.

39. Novak, P., Cabelka, J. (1981). *Models in hydraulic engineering.* Pitman, Boston.

40. Novak, P., Moffat, I., Nalluri, C., Narayanan, R. (2006). *Hydraulic structures*, ed. 4. CRC-Press, Boca Raton, FL.

41. Novak, P., Guinot, V., Jeffrey A., Reeve, D.E. (2010). *Hydraulic modeling: An introduction.* Taylor & Francis: Oxford.

42. Olsen, N.R.B., Skoglund, M. (1994). Three-dimensional numerical modeling of water and sediment flow in a sand trap. *J. Hydraulic Res.* 32(6), 833–844. doi: 10.1080/00221689409498693 ,

43. Padhy, M.K., Saini, R.P. (2009). Effect of size and concentration of silt particles on erosion of Pelton turbine buckets. *Energy*34(10), 1477–1483. doi: 10.1016/j.energy.2009.06.015 ,

44. Padmanabhan, M. (1984). Air ingestion due to free-surface vortices. *J. Hydraulic Eng.* 110(12), 1855–1859. doi:10.1061/(ASCE)0733-9429(1984)110:12(1855)

45. Pfister, M., Hager, W.H. (2010). Chute aerators II: Hydraulic design. *J. Hydraulic Eng.* 136(6), 360–367. doi:10.1061/(ASCE)HY.1943-7900.0000201 ,

46. Pfister, M., Lucas, J., Hager, W.H. (2011). Chute aerators: Preaerated approach flow. *J. Hydraulic Eng.* 137(11), 1452–1461. doi: 10.1061/(ASCE)HY.1943-7900.0000417 ,

47. Pfister, M., Hager, W.H., Boes, R.M. (2014). Trajectories and air flow features of ski jump-generated jets. *J. Hydraulic Res.*52(3), 336–346. ,

48. Ranga Raju, K.G., Kothyari, U.C., Srivastav, S., Saxena, M. (1999). Sediment removal efficiency of settling basins.. *J. Irrig. Drain. Eng.* 125(5), 308–314. doi: 10.1061/(ASCE)0733-9437(1999)125:5(308) , ,

49. Raynal, S., Courret, D., Larinier, M., David, L. (2013). An experimental study on fish-friendly trashracks 2: Angled trashracks.*J. Hydraulic Res.* 51(1), 67–75. doi: 10.1080/00221686.2012.753647 ,

50. Sagar, B.T.A. (1995). ASCE Hydrogates Task Committee design guidelines for high-head gates. *J. Hydraulic Eng.* 121(12), 845–852. doi: 10.1061/(ASCE)0733-9429(1995)121:12(845) ,

51. Schmocker, L., Pfister, M., Hager, W.H., Minor, H.-E. (2008). Aeration

characteristics of ski jump jets. *J. Hydraulic Eng.* 134(1), 90–97. doi: 10.1061/(ASCE)0733-9429(2008)134:1(90) ,

52. Schnitter, N.J. (1994). *A history of dams.* Balkema, Rotterdam.

53. Schröder, R.C.M. (1990). Hydraulische Methoden zur Erfassung von Rauheiten (Hydraulic methods to determine roughnesses). In *DVWK Schrift* 92. Parey, Hamburg and Berlin (in German).

54. Sharma, H.R. (1976). Air entrainment in high-head gated conduits. *J. Hydraulics Div.* ASCE 102(HY11), 1629–1646; 103(HY10), 1254–1255; 103(HY11), 1365–1366; 103(HY12), 1486–1493; 104(HY8), 1200–1202.

55. Speerli, J., Hager, W.H. (2000). Air flow characteristics in bottom outlets. *Can. J. Civil Eng.* 27(3), 454–462. doi: 10.1139/l99-087 ,

56. Steiner, R., Heller, V., Hager, W.H., Minor, H.-E. (2008). Deflector ski jump hydraulics. *J. Hydraulic Eng.* 134(5), 562–571. doi:10.1061/ (ASCE)0733-9429(2008)134:5(562) ,

57. Suerich-Gulick, F., Gaskin, S., Villeneuve, M., Parkinson, E. (2014). Characteristics of free-surface vortices at low-head hydropower intakes. *J. Hydraulic Eng.* 140(3), 291–299. doi: 10.1061/(ASCE)HY.1943-7900.0000826 ,

58. Sumi, T., Kantoush, S.A. (2011). Comprehensive sediment management strategies in Japan: Sediment bypass tunnels. Proc. 34th *IAHR World Congress*, Brisbane, 1803–1810.

59. Sumi, T., Okano, M., Takata, Y. (2004). Reservoir sedimentation management with bypass tunnels in Japan. Proc. 9th Intl. Symp. *River Sedimentation*, Yichang PRC, 1036–1043.

60. Vischer, D.L., Hager, W.H. (1998). *Dam hydraulics.* Wiley, Chichester.

61. Wagner, W.E. (1967). Glen Canyon diversion tunnel outlets. *J. Hydraulics Div.* ASCE 93(HY6), 113–134.

62. Wood, I.R. (1991). Air entrainment in free surface flows. In *IAHR Hydraulic Structures Design Manual* 4. Balkema, Rotterdam.

63. Yıldırım, N., Akay, H., Taştan, K. (2011). Critical submergence for multiple pipe intakes by potential solution. *J. Hydraulic Res.* 49(1), 117–121. doi: 10.1080/00221686.2010.535651

Chapter 8

UNSATURATED HYDRAULIC CONDUCTIVITY OF FRACTAL-TEXTURED SOILS

Yongfu Xu

Department of Civil Engineering, Shanghai Jiao Tong University, Shanghai, China

INTRODUCTION

The increasing concern with groundwater pollution and contamination of soils has stimulated the development of numerous mathematical models of pollutant transport in soils. The most important approaches to model transient water and solute transport in the vadose zone are based on the Richards equation. To solve this equation, the knowledge of the soil hydraulic properties, namely, the soil-water characteristic curve (SWCC) and the unsaturated hydraulic conductivity is required. The laboratory measurements show that the value of unsaturated hydraulic conductivity varies considerably from soil to soil with different water content (Khaleel and Relyea, 1995). Indeed, it is found that the unsaturated hydraulic conductivity decreases by one to three orders of magnitude across a small pressure head range even near saturation (0~10cm pressure head), due to the effects of structural macropores (Jarvis and Messing, 1995). Of all hydraulic properties, the unsaturated hydraulic conductivity is most difficult to measure. Therefore the use of indirect methods has become more and more common to estimate the unsaturated hydraulic conductivity from more easily measured soil properties (van Genuchten et al., 1992).

Modern hydrological models require information on hydraulic conductivity and soil-water retention characteristics. All hydraulic properties, the soil-water characteristics, hydraulic conductivity and soil-water diffusivity (SWD) are closely related to the geometry of a porous media (Brooks and Corey, 1966; Burdine, 1953). Measurements of hydraulic properties are expensive, time-consuming and highly variable (Dirksen, 1991). Prediction of these properties is a viable alternative, especially when the predictive model contains a few parameters sensitive to structural conditions. Porous media (e.g. soils, rocks, etc.) are heterogeneous systems composed of numerous, different and interacting components (van Damme, 1995). The complex nature of these porous media complicates any prediction of their hydraulic properties.

A potentially powerful method results from the pore surface models, in which the soil–water characteristic curve of unsaturated soils is interpreted as statistical measure of its equivalent pore-size distribution (PSD) (van Genuchten, 1980; Corey, 1992). The frequency of different pore radii is related to matric suction by the soil-water characteristic curve (SWCC) and the relation between matric suction and pore radius is described by the Young-Laplace equation. The relative hydraulic conductivity (RHC) of unsaturated soils can be deduced from the soil-water characteristic curve (SWCC) through simplifying assumptions on pore topology and using the Hagen-Poiseuille law as an approximation for water flow. An additional empirical parameter is introduced, which includes all uncertainties. This empirical parameter is often referred to as "tortuosity", but its physical meaning is unclear (Vogel and Roth, 1998). The choice of the analytical model for the soil-water characteristic curve (SWCC) can significantly affect the predicted function of the unsaturated hydraulic conductivity.

Fractals describe hierarchical systems and are suitable to model the heterogeneous soil structure with tortuous pore space (Rieu and Sposito,1991; Xu and Sun, 2002). Toledo et al. (1990) modeled the soil-water characteristic curve (SWCC) and unsaturated hydraulic conductivity using fractal geometry and thin-film physics. Tyler and Wheatcraft (1990) gave the unsaturated hydraulic conductivity functions based on the fractal model for the soil-water characteristic curve (SWCC) and the relative conductive models developed by Mualem (1976) and Burdine (1953). Crawford (1994) studied the influence of heterogeneity of both the solid matrix and the pore space, as well as the shape of the pore boundary, on the saturated and unsaturated hydraulic conductivity. Fuentes et al. (1996) derived an expression for unsaturated hydraulic conductivity using the fractal dimension obtained from the soil-water characteristic curve (SWCC). Hunt and Gee (2002) applied critical path analysis from percolation theory to calculate the unsaturated hydraulic conductivity of soils with fractal pore space. Xu (2004),Xu and Dong (2004) Xu et al. (2004) derived the unsaturated hydraulic conductivity using the fractal model for the pore surface and gave a simple method to determine the fractal dimension. Models of the unsaturated hydraulic conductivity incorporate fractal dimension characterizing scaling of different properties including parameters representing connectivity and tortuosity. Thus, it is encouraged to derive the functions of the soil–water characteristic curve and unsaturated hydraulic conductivity from the fractal model for the pore surface. In this chapter, the soil-water characteristic curve (SWCC) and relative hydraulic conductivity (RHC) function were derived and expressed by the effective degree of saturation based on the fractal model for the pore surface. The proposed soil–water characteristic curve and unsaturated hydraulic conductivity function were examined in detail by the published experimental data. Comparisons between the prediction using both the fractal model and the van Genuchten-Mualem (G-M) model and the measurement of the relative hydraulic conductivity (RHC) were conducted. Using the fractal model for the soil-water characteristic curve (SWCC) and relatively hydraulic conductivity (RHC), one-dimensional rainfall infiltration and slope stability due to rainfall infiltration were analysized in chapter.

FRACTALS

Some of the most common and useful examples of fractals are: the Koch curve, the Sierpinski gasket and carpet, and the Menger sponge. All of these fractals enjoy the property of self-similarity. Roughly speaking, a subset of $\Re^n$ is said to be self-similar if it is a union of a number of smaller similar copies of itself. Some of the essential notions of the theory of fractals as it applies to self-similar sets should be briefly reviewed. A previous acquaintance with at least one of the sets mentioned above is useful but not necessary. Given an arbitrary subset B of $\Re^n$, a cover of B is a family U of sets $U0 \subset \Re^n$ such that:

$$B \subset \bigcup_n U_\alpha$$

(1)

If the family U is countable (or finite) the cover is said to be countable (or finite). It is customary in that case to indicate the members of the family with a Latin subscript, namely, U_i, where i ranges over the natural numbers. The diameter of a subset of $\Re^n$ is the least upper bound (i.e., the supremum) of the distance between pairs of points in the subset. By convention, the diameter of the empty set is zero. A δ-cover of $B \subset \Re^n$ is a (countable) cover such that, for every i, $\text{diam}(U_i) \leq \delta$, where δ is a positive real number and where diam(.) is the diameter function on subsets of $\Re^n$. The members of a δ-cover are indicated by U_i^δ. The s-dimensional Hausdorff measure $H_s(B)$ of B is defined as:

$$H^s(B) = \lim_{s \to 0} \inf \sum_{i=1}^{n} [\text{diam}(U_i^s)]^s$$

(2)

where the inf extends over all δ-covers. It can be shown that this definition indeed provides an outer measure for all non-negative values of s and for any subset of $\Re^n$. The Hausdorff dimension of B is defined as:

$$\dim_H B = \inf\{s \geq 0 : H^s(B) = 0\}$$

(3)

The Hausdorff dimension of a subset of $\Re^n$ cannot exceed n. If $\dim_H B > 0$, then it can be shown that for all values of s strictly smaller (respectively, larger) than the dimension, the s-dimensional Hausdorff measure of B is infinite (respectively, zero). In this sense, the Hausdorff dimension represents a critical value of discontinuity of the s-dimensional Hausdorff measure, thus making Eq. (3) meaningful. Of particular interest is the value of the s-dimensional Hausdorff measure for s equal to the Hausdorff dimension of the set. This is usually called the Hausdorff measure of the set. The Hausdorff measure of a set may turn out to have any value, including zero or infinite.

An important feature of the s-dimensional Hausdorff measure is the scaling property, namely, the way it changes under similarity transformations of $\Re^n$. Recall that a transformation S: $\Re^n \to \Re^n$ is called a similarity if there exists a real scale factor $\lambda > 0$ such that, for all x, y $\in \Re^n$, the following equation is satisfied:

$$\left|S(y) - S(x)\right| = \lambda\left|y - x\right|$$

(4)

where |.| denotes the Euclidean distance in $\Re n$. It is not difficult to prove that for all subsets $B \subset \Re^n$ and for all values of s, under a similarity transformation S with scale factor λ the s-dimensional Hausdorff measure transforms according to the formula:

$$H^s(S(B)) = \lambda^s H^s(B)$$

(5)

Another important property of the s-dimensional Hausdorff measure is that it is preserved under Euclidean isometries (translations, rotations, reflections). In other words, congruent sets have the same s-dimensional Hausdorff measure. The behavior of the Hausdorff measure under general affine transformations, on the other hand, cannot be captured under the umbrella of a simple formula.

The properties just described of the s-dimensional Hausdorff measure can be used to obtain a straightforward evaluation of the Hausdorff dimension of self-similar fractals. Indeed, let B be the union of m copies of λ-scaled mutually congruent copies of itself (with $\lambda < 1$). If these copies are disjoint Borel sets, we have:

$$H^s(B) = m\lambda^s H^s(B)$$

(6)

by virtue of the scaling property (5). Assume now that the Hausdorff measure of B (namely, the value of $H_s(B)$ for $s=\dim H(B)$) is finite and positive. In that case, it follows Eq. (6) that:

$$\dim_H B = -\frac{\log m}{\log \lambda}$$

(7)

The fractal dimension for a pore surface can be defined in the following way. Imagine the pores being enclosed by a set of spheres of radius r, the number of spheres N necessary to do this is clearly a function of radius of the spheres. The definition of the fractal dimension D is:

$$D = \lim_{r \to 0}\left[\frac{\ln(N(A,r))}{\ln(1/r)}\right]$$

(8)

Equation (8) is often used to determine the fractal dimension by experiments. A typical fractal set, the middle third Cantor set may be constructed from a unit interval by a sequence of deletion operations (Fig. 1). Let E_0 be the interval [0,1], E_1 is the set obtained by deleting the middle third of E_0 so E_1 consists of the two intervals $\left[0, \frac{1}{3}\right]\left[\frac{2}{3}, 1\right]$. Deleting the middle third of E_1 gives E_2, thus E_2 comprises the four intervals $\left[0, \frac{1}{9}\right]\left[\frac{2}{9}, \frac{3}{9}\right]\left[\frac{6}{9}, \frac{7}{9}\right]\left[\frac{8}{9}, \frac{9}{9}\right]$. Proceeding in this like manner, E_i is obtained by deleting the middle third of each interval in E_{i-1} and Ei consists of 2^i intervals with length of 3^{-i}. Using Eq. (8), the fractal dimension of the middle-third Cantor set is $D = \lim_{i \to \infty}\left[\frac{\log(2^i)}{\log(1/(3^{-i}))}\right] = 0.63$. The Cantor set is often used to model the dust distribution.

Figure 1: Middle-third Cantor set with fractal dimension D=0.63

FRACTAL MODEL FOR THE PORE SURFACE

Many research results show that the soil pore surface is fractal (Avnir and Jaroniec 1989; Xu and Sun 2002). To investigate the impacts of fractal scaling upon hydraulic properties of porous media, a fractal representation of a porous media is developed. A cube of size 1 by 1 is formed by initially subdividing the cube into $1/a^3$ subcubes, each of size a by a. From the original cube, N subcubes are removed and represent a pore of size a by a. The remaining 1/

a^3-N subcubes are then each divided into $1/a^3$ subcubes and N subcubes are removed from each of the original subcubes. Such a recursion algorithm results in a cube everywhere filled with pores of all sizes, with a predominance of small pores. The number of size of a^3 needed to cover the pores equal to or larger than a^3 is given by N. The Menger sponge (Fig. 2) is often used to model porous materials, such as soils. For the Menger sponge, a=1/3, N=7, and the fractal dimension is given by

$$D = \lim_{i \to \infty} \left\{ \frac{\log(20^i)}{\log(1/(3^{-i}))} \right\} = 2.73.$$

Figure 2: The Menger sponge with fractal dimension D=2.73

When we covered a soil pore surface using balls with the same radius, the soil pore surface is covered by N_1 balls with radius r_1, the same surface is covered by N_i balls with radius r_i, and it is covered by N_j balls with radius r_j, and usually $r_1 > r_i > r_j$, it is found that the surface area of the soil pore surface contents has the following relationship: $N_0 r_0^2 < N_i r_i^2 < N_j r_j^2$. If the soil pore surface is a self-similar surface, i.e., the pore surface is a fractal surface, the relationship between the number of covered balls and its radius can be written as (Mandelbrot 1982)

$$N(r) = Cr^{-D} \qquad (9)$$

where C is a constant, r is the size corresponding to the pore radius, D is the fractal dimension of the pore surface. All materials have a surface fractal dimension in the whole range of physically meaningful values, i.e., from 2 to

3. The limit D=2 would correspond to a perfectly regular, smooth (Euclidean) surface, whereas D=3 would correspond to a self-similar surface so intricate that it would be space filling.

The fractal dimension of the pore surface can be determined using mercury intrusion porosimetry and adsorption isotherm. These two methods are introduced in detail as follows.

Mercury Intrusion Porosimetry

The d-measure is obtained from Eq. (9), and is expressed as follows:

$$M(r) = N(r)r^d = Cr^{d-D}$$
(10)

For the fractal pore surface, the relationship between the surface area with the radius less than r and the pore radius r can be obtained from Eq. (10). The surface area can written as

$$A_p(\leq r) = Cr^{2-D} \qquad a)$$
$$V_p(\leq r) = Cr^{3-D} \qquad b)$$
(11)

where in 11 a) $A_p(\leq r)$ is the surface area of the soil pores with the radius smaller than r. Similarly, the total volume of the pores with the radius less than r is given by 11b), where in $V_p(\leq r)$ is the cumulative volume of the soil pores with the radius smaller than r.

Neimark (1992) gave a method to measure the surface fractal dimension of the soil pores using mercury intrusion porosimetry. The soil pore surface can be approximated by the equilibrium interface between mercury and soil particles in the close vicinity of the pore surface. According to the Young–Laplace equation, the mean radius of the equilibrium interface is expressed as follows:

$$r(p) = \frac{2\sigma \cos \alpha}{p}$$
(12)

where r(p) is the mean radius of the equilibrium interface under pressure p, σ is surface tension. From the thermodynamic viewpoint, the equilibrium interface area can be calculated from the balance between the work of formation of the equilibrium interface and the work of mercury intrusion (Neimark 1992), i.e.,

$$A_p(p) = \int_0^{V(p)} \frac{V(p)}{T}dp$$
(13)

where $A_p(p)$ is the equilibrium interface area under pressure p. The surface fractal dimension can be determined theoretically from Eq. (11a). If the slope of the straight line is λ in log r(p) versus log A_p(p), the surface fractal dimension of soil pores is given by

$$D = 2 - \lambda \qquad (14)$$

The pore-size distribution (PSD) is usually obtained using mercury intrusion porosimetry. According toEq.(11b), the fractal dimension of the pore surface can be obtained from the slope of the regressed linear relation in the plane of logr vs. logV_p. If the slope of the fitting line in the logr-logV_p plane is κ, the fractal dimension of the pore surface is written as

$$D = 3 - \kappa \qquad (15)$$

The value of fractal dimension spans a large range from 1.0 to 3.0 (Gimenez et al., 1997). Larger fractal dimension is associated with clayey soils (van Damme, 1995).

The accumulative volume of the pores can be measured by many methods. Mercury intrusion porosimetry provides a useful and potentially valuable measurement of the pore volume for the porous medium. The ranges of equivalent pore diameter explored cover almost five orders of magnitude, from several hundred microns down to approximately 16Å (Watabe et al., 2000).

Figure 3 shows the soil pore-size distribution (PSD) obtained from mercury intrusion porosimetry (Watabe et al., 2000). Symbols S-02, S-03 and S-04 are the serial numbers of soil samples, and V_v is the total pore volume in Fig. 3. It is seen from Fig. 3 that the surface of the soil pores can be described by fractal model, and the relationship between V_p($\leq$r) and r can be expressed by a linear function in log-log plot. The slopes of the regressed linear relation in the plane of logr vs. log(V_p($\leq$r)/V_v) are 0.34, 0.37 and 0.49, and therefore the fractal dimensions are 2.66, 2.63 and 2.51 for specimens of S-02, S-03 and S-04, respectively. The maximum radius is the radius at which the pore volume reaches the maximum value, and is defined as the intersection between the fitting line and the line of V_p($\leq$r)/V_v=100%. The maximum radii of soil pores are 0.0125, 0.06 and 0.006 mm for specimens of S-02, S-03 and S-04, respectively. The relationship between the maximum radius R and the air-entry value ψ_e can be expressed by the Young-Laplace equation, i.e. ψ_e=2σcosα/R. The parameters α=0 and σ=0.075kPa mm (Watabe et al., 2000), and the air-entry values are 12kPa, 2.5kPa and 25kPa corresponding to the maximum pore radius R of 0.0125, 0.06 and 0.006 mm for specimens of S-02, S-03 and S-04, respectively. The parameters obtained from the fractal model of the pore surface are listed in Table 1. InTable 1, parameters R and κ are obtained from the fractal model of the pore surface in Fig. 3.

Table 1: Parameters obtained from the fractal model of PSD

Soil type	κ	D	δ	*R (mm)*	$\psi_e(kPa)$	**Data source**
S-02	0.34	2.66	-0.34	0.0125	12	Watabe et al., 2000
S-03	0.37	2.63	-0.37	0.06	2.5	
S-04	0.49	2.51	-0.49	0.006	25	
A horizon	0.22	2.78	-0.22	0.2	0.75	Vogel and Roth, 1998
B horizon	0.13	2.87	-0.13	0.25	0.6	
Toyoura sand	1.33	1.67	-1.33	0.09	1.67	Uno et al., 1998

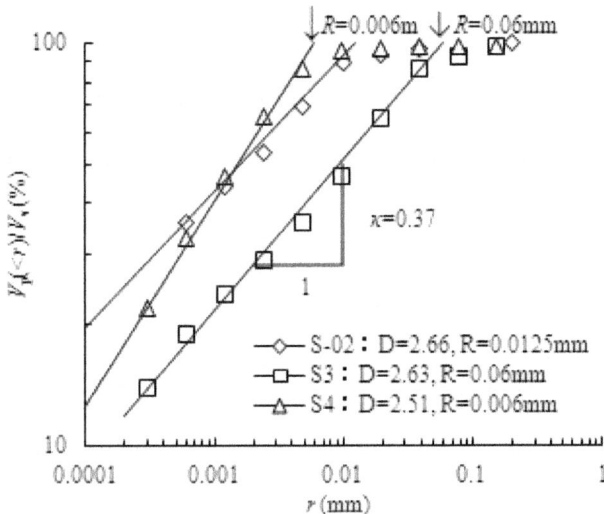

Figure 3: Fractal dimension and the maximum pore radius obtained from PSD for glacial tills (Data fromWatabe et al., 2000)

Experimental data of the pore-size distribution (PSD) for agricultural silty soils from two horizons, A and B were also taken from Vogel and Roth (1998). The pore-size distribution (PSD) was determined using the three-dimensional reconstructions of the pore space. The pore space was eroded by a spherical structuring element with a given diameter 2r and subsequently dilated by the same structuring element (Vogel and Roth, 1998). In the resulting image all pores smaller than 2r were removed. The cumulative volume $V_p(>r)/V_T$ was obtained using the application of successively larger structuring elements byVogel and Roth (1998). The following relationship is obtained between $V_p(>r)/V_T$ and $V_p(\leq r)/V_T$.

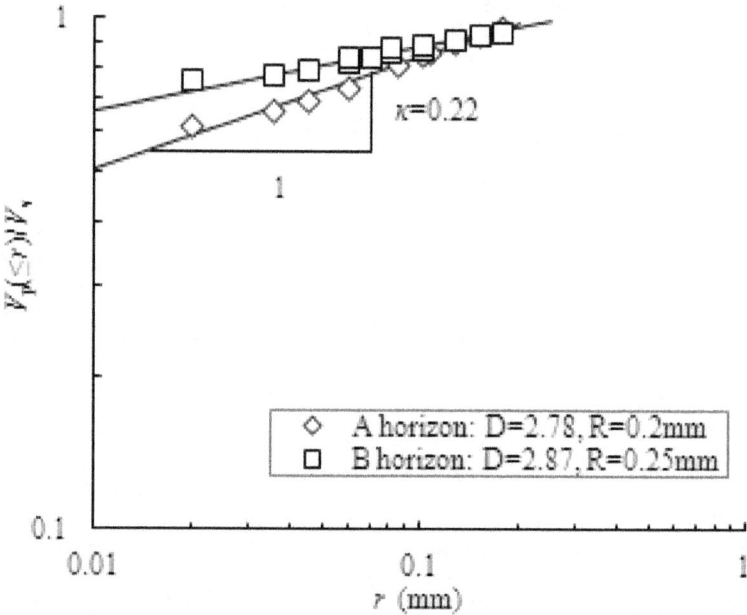

Figure 4: Fig. 4 Fractal dimension and the maximum radius of the PSD of two silty soils (Data from Vogel and Roth, 1998)

$$\frac{V(>r)}{V_T} + \frac{V(\leq r)}{V_T} = \frac{V_v}{V_T} = \theta_s \tag{16}$$

where V_T is the total volume of soil, V_v is the total volume of void, θ_s is the saturated volumetric water content. The saturated volumetric water contents are 0.502 and 0.448 for the soils at A horizon and B horizon, respectively. Hence the cumulative volume $V(\leq r)/V_T$ can be obtained from the cumulative volume $V(>r)/V_T$. The cumulative volume of $V(\leq r)/V_T$ can be translated into $V(\leq r)/V_v$ according to Eq.(16). The cumulative volume of $V(\leq r)/V_v$ is shown in Fig. 4, where the dots denote the results transformed from the experimental data given by Vogel and Roth (1998) and the solid lines denote the results of the fractal model. The relationship between the pore volume and the pore radius satisfies an approximately linear expression in log-log plot. The slopes (κ) of the regression line are 0.24 and 0.13 for the soils at A and B horizons, respectively. According to Eq. (15), the fractal dimensions of the pore surface are 2.76 and 2.87 for A and B horizons, respectively. The maximum radius is defined as the value at which the cumulative volume of pores reaches the maximum, and $V(\leq r)/V_v = 100\%$. The maximum radii of the soil pores at the A and B horizons are 0.2cm and 0.25cm, respectively. The air-entry values can be calculated using the Young-Laplace equation, and they are 0.75kPa and

0.6kPa for the A and B horizons, respectively. The fractal dimension of the soil pore and the air-entry value of the soils at the A and B horizons are listed in Table 1. It can be seen from Figs. 3-4 that the pore surface can be modeled by fractal theory, and the pore-size distribution (PSD) satisfies with Eq. (11b). The soil structure can be described by the fractal dimension. The larger the fractal dimension, the more tortuous is the pore space. The fractal dimension is varied with the soil structure, and is different for different soils.

Uno et al. (1998) studied the relationship between the pore-size distribution (PSD) and unsaturated hydraulic properties of Toyoura sand. The pore-size distribution (PSD) of Toyoura sand was measured by the air intrusion method, and is shown in Fig. 5. The relationship between the pore volume and pore radius is approximated by a linear function in log-log plot, and satisfies with Eq. (11b). Hence the pore-size distribution (PSD) of Toyoura sand can be described by fractal model, and its fractal dimension is 1.67 obtained from Fig. 5 not increase with radius, i.e., $V_p/V_v=1$. The maximum pore radius of Toyoura sand is nearly 0.09 mm in Fig. 5. The surface tension (T) of water is 0.075kPa mm. The air-entry value can be calculated from Young-Laplace equation, and is 1.67kPa on the assumption that $\alpha=0$. The fractal dimension of the soil pore and the air-entry value of Toyoura sand arc listed in Table 1.

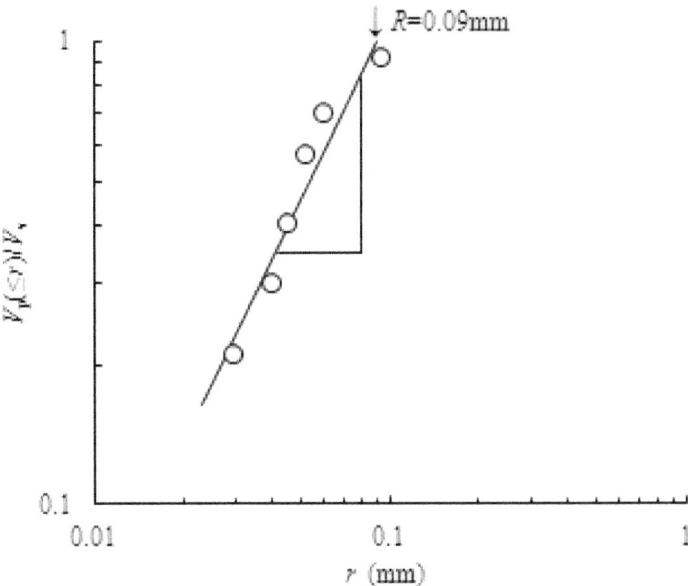

Figure 5: The fractal dimension and the maximum radius evaluated from the PSD of Toyoura sand. (Data fromUno et al., 1998.)

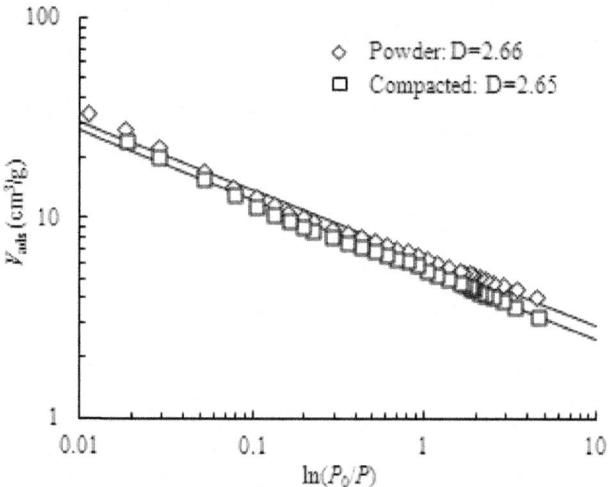

Figure 6: The surface fractal dimension of Tsukinuno bentonite from nitrogen adsorption

Adsorption Isotherm

The fractal dimension of the pore surface serves to characterize the pore surface roughness or irregularities. The magnitude of the fractal dimensions is relevant to many important physicochemical processes namely adsorption, surface diffusion and catalysis. Avnir and Jaroniec (1989) proposed a convenient method to determine the fractal dimension from the nitrogen adsorption. Similarly with the adsorption isotherm equation, the correlation of the water volume absorbed by clay to the vapour pressure is given by

$$\frac{V_w}{V_m} = k\left[\ln\left(\frac{P_0}{P}\right)\right]^{D-3}$$

(17)

where V_w is the water volume absorbed by clay, V_m is the volume of montmorillonite, P is the partial water vapour pressure in equilibrium with clay at some water content w and some temperature T and P_0 is the equilibrium water vapour pressure of pure water at temperature T. The surface fractal dimension of Tsukinuno bentonite obtained from nitrogen adsorption is shown in Fig. 6. The surface fractal dimensions of bentonite powder and compacted bentonite are 2.66 and 2.65, respectively.

The adsorption isotherm also allows the calculation of swelling pressure of clay as a function of the water content (Kahr et al., 1990). Satisfactory agreement is found between the calculated and the experimental data for water content bigger than 10%.

$$p_s = -\frac{\bar{R}T}{M_w v_w} \ln\left(\frac{P_0}{P}\right)$$

$$(18)$$

where p_s is swelling pressure, the maximum axial pressure which is needed to maintain the original sample height, $\bar{R}$ is the molar gas constant, T is the Kelvin temperature, M_w is molecular mass of water and v_w is partial specific volume of water. Combining Eq. (17) with (18), the relationship between normalized water volume by the clay volume and swelling pressure is written as

$$\frac{V_w}{V_m} \propto p_s^{D-3}$$

$$(19)$$

The surface fractal dimension of clay can be obtained from Eq. (19). The experimental data of the swelling pressure and swelling deformation tests are compiled in Fig.7. It is seen that the relationship between the normalized water content and swelling pressure or vertical overburden pressure can be described by the same linear function which is expressed in Eq. (19). Therefore, Eq. (19) represents a general relationship between the water volume absorbed by clay per unit volume and vertical pressure supplied to specimen in swelling process. The surface fractal dimension of Tsukinuno bentonite is 2.63, which nearly equals to that obtained from Fig. 6. In Fig. 6, C_b is the bentonite content.

Swelling deformation tests of Wyoming bentonite were also conducted by Millins et al. (1996) and Studds et al. (1998). The relationship between the clay void ratio (e_c) and vertical pressure is shown inFig. 8. Relationship between the clay void ratio (e_c) and normalized water volume is given by

$$e_c = \frac{1}{C_m}\frac{V_w}{V_m}$$

$$(20)$$

where C_m is the montmorillonite content, and C_m=75%. It is seen that the slope of $\log e_c$-$\log p$ equals to that of $\log(V_w/V_m)$-$\log p$ from Eq. (20) for C_m being constant. Thus, the surface fractal dimension of Wyoming bentonite can be obtained from the linear relationship of $\log e_c$-$\log p$. The surface fractal dimensions of Wyoming bentonite are 2.64, which obtained from the swelling deformation tests conducted by different authors.

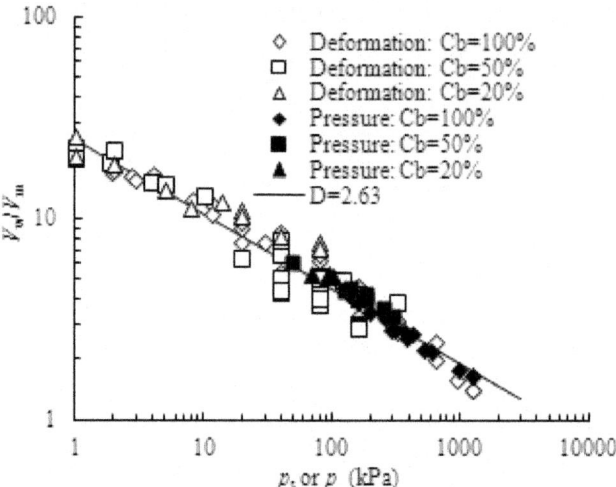

Figure 7: Relationship between absobed water volume and swelling pressure or vertical overburden pressure of Tsukinuno bentonite

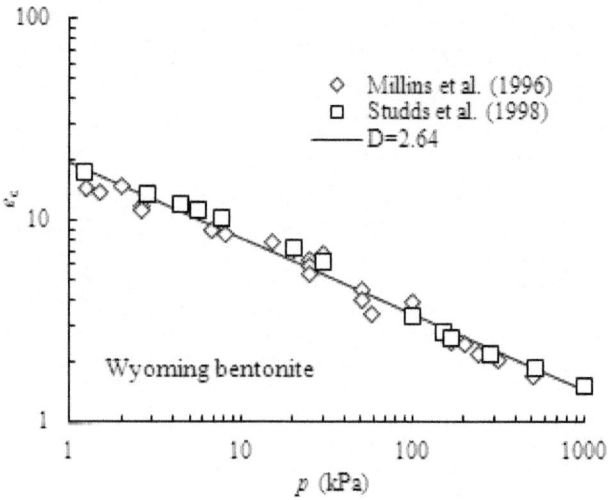

Figure 8: Relationship between clay void ratio and vertical overburden pressure of Wyoming bentonite

Correlation between the Fractal Dimension of PSD and that of the GSD

The pore-size distribution (PSD) is not often and easily measured experimentally. The grain-size distribution (GSD) is easily measured and the fractal dimension of the pore surface can be conveniently obtained from the grain-size distribution (GSD). The relationship between the pore-size distribution (PSD) and the grain-size distribution (GSD) can be constructed for the soils with a homogeneous fabric. Consider a pore limited by a number of grains N_g, the pore radius r is related to the radius r_g of the smallest grain as follows (Watabe et al., 2000)

$$r = f_p r_g$$

(21)

where fp is a constant less than 1.0. If the particles are randomly arranged, the probability of having a grain of radius less than r_g is dP_g, the probability of having the N_g-1 grains with a radius greater than r_g is $(1-P_g)^{N_g-1}$. The probability P of having a pore with a radius less than r is given by (Watabe et al., 2000)

$$dP = N_g \left(1 - P_g\right)^{N_g - 1} dP_g$$

(22)

It is assumed that the grain-size distribution (GSD) satisfies the fractal model, and is given by

$$P_g = \left(r_g / R_g\right)^{3 - D_g}$$

(23)

where D_g is the fractal dimension of the grain-size distribution (GSD), R_g is the maximum radius of soil grains. The probability P is given by (Watabe et al., 2000)

$$P = 1 - \left[1 - \left(\frac{r}{f_p R_g}\right)^{3 - D_g}\right]^{N_g}$$

(24)

If $r/(f_g R_g)$<<1, Eq.(24) can be written as

$$P = N_g \left(\frac{r}{f_p R_g}\right)^{3 - D_g}$$

(25)

It is seen that the pore surface has the same fractal dimension as that obtained form the grain-size distribution (GSD) from Eqs. (23) and (24). Comparisons between the pore-size distribution (PSD) and the grain-size distribution (GSD) are shown in Fig. 9, where $P=V(\leq r)/V_v$ and $P_g=M(\leq r)/M_T$, V and Mare the pore volume and grain mass, respectively. It is seen from Fig. 9 that both the pore-size distribution (PSD) and the grain-size distribution (GSD) can be expressed by fractal model, and they have the nearly same fractal dimension.

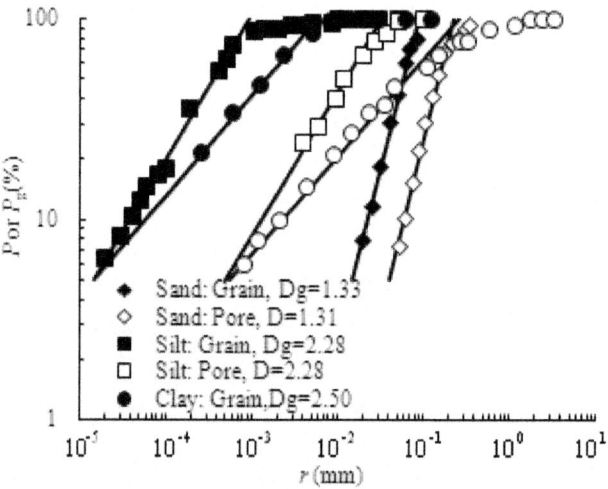

Figure 9: Correlation of the pore-size distribution (PSD) to the grain-size distribution (GSD).

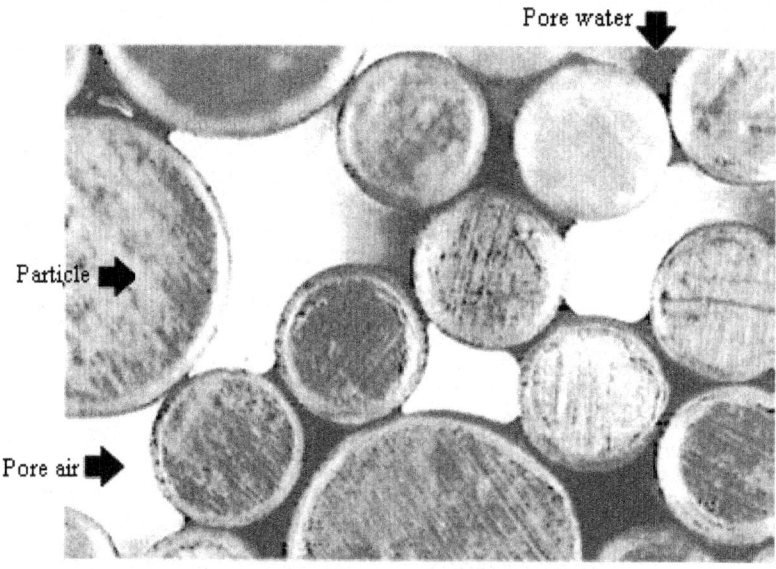

Figure 10: Model test for simulating the water distribution

SOIL-WATER CHARACTERISTIC CURVES (SWCC)

The soil-water characteristic curve defines the relationship between water content (gravimetric water content w, volumetric water content θ) or degree of saturation (S) and matric suction (ψ). The soil water characteristic curve (SWCC) has been used extensively in the study of unsaturated soils and relates the water content and soil matric suction. The relationship between the matric suction and the radius of the incurvated surface between the pore-air and pore-water is expressed by the Young-Laplace equation. The hydraulic and mechanical properties of unsaturated soils are correlated to the microstructures of soils through the Young-Laplace equation.

For unsaturated soils, the pore water is usually localized in small micropores, and the micropores are usually fully filled with water. The water content in macropores is very little, sometimes, there is no water in macropores in highly unsaturated soils (Fig. 10). The water distribution (black domain) of model test shown in Fig. 10 proved the above assumptions. In the model test, the soil grains of different sizes were simulated by the aluminum rods with different radius (Matsuoka, 1999). The water mainly distributed in the small pore and no water exists in the large pore for unsaturated soils in the model test.

Three air-water distribution states were identified as saturated funicular state, complete pendular state, and partial pendular state, respectively. All the voids are filled with liquid in the saturated state. Air bubbles are present in the voids in the funicular state, and the liquid phase is continuous. There is no continuity in the water phase in the pendular state. Pore water within unsaturated soils can be divided into three forms (Wheeler and Karube, 1996) (Fig. 11): bulk water within those void spaces that are completely flooded, meniscus water surrounding all inter-particle contact points that are not covered by bulk water and absorbed water (which is tightly bounded to the soil particles and acts as parts of the soil skeleton). The relationship between the soil-water characteristic curve (SWCC) and the pore-water forms is shown in Fig. 11. When the volumetric water content is less than its residual value, the pore-water is tightly absorbed by soil particle and cannot move freely. This absorbed water is seen as a part of soil particles. Thus, the contribution of the filled pores with radius r→dr to the water content is given by

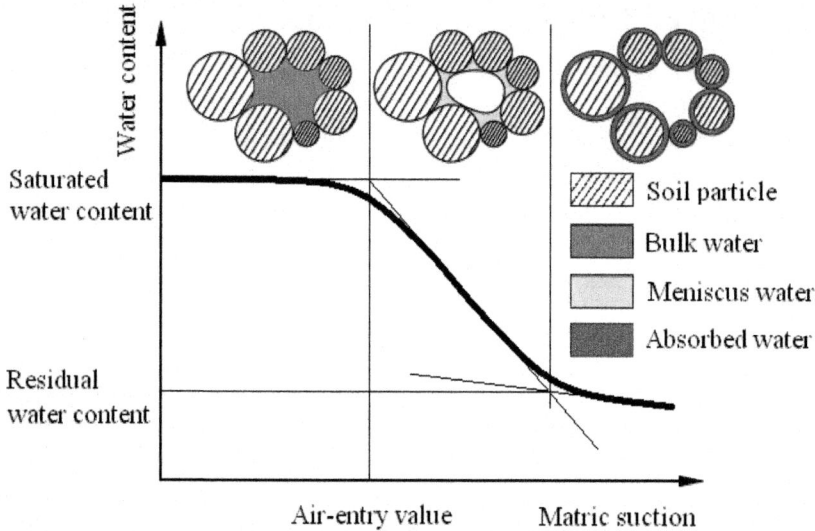

Figure 11: Relationship between the SWCCs and the pore-water forms of unsaturated soil (Matsuoka, 1999)

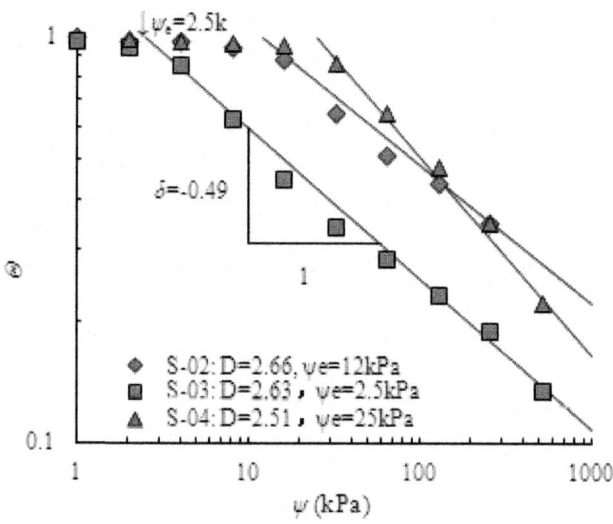

Figure 12: Comparison between the prediction and experiments of the SWCCs (Data from Watabe et al., 2000)

$$dΛ = \frac{N4\pi r^2 dr}{V_T}$$

(26)

where Λ is the relative volumetric water content, and $\Lambda=\theta-\theta_r$, θ and θ_r are the actual and residual volumetric water content, respectively. The residual volumetric water content is the volumetric water content at which the effectiveness of matric suction to cause further removal of water requires vapour migration. Substituting Eq. (9) into Eq. (26), Λ is given by

$$\Lambda = Br^{3-D}$$
(27)

where $B=4\pi C/[V_T(3\text{-}D)]$. Similarly with Eq. (27), the relatively volumetric water content at saturation is written as follows

$$\Lambda_s = BR^{3-D}$$
(28)

where R is the maximum radius of soil pores.

The relationship between matric suction ψ and the pore radius is obtained from the Young-Laplace equation. Thus, the soil-water characteristic curve (SWCC) is obtained as follows

$$\Theta = \left(\frac{\psi}{\psi_e}\right)^{\delta}$$
(29)

where $\delta=D\text{-}3$, Θ is the normollized volumetric water content, and is written as follows:

$$\Theta = \frac{\theta-\theta_r}{\theta_s-\theta_r} = \frac{S-S_r}{100-S_r}$$
(30)

where θ, θ_r and θ_s are the volumetric water content, residual volumetric water content and saturated volumetric water content, S and S_r are the degree of saturation and residual degree of saturation. The residual volumetric water content is not always made routinely, in which case it has to be estimated by extrapolating available soil-water characteristics data towards lower water content, such as shown inFig. 11. van Genuchten (1980) defined the residual volumetric water content as the water content for which the gradient $d\theta/d\psi$ becomes zero at high matric suction. From a practical point of view it seemed sufficient to define θ_r as the water content at some large matric suction, e.g. at the permanent wilting point $\psi_e=1500$kPa.

Equation (29) is the soil-water characteristic curve (SWCC) derived from the fractal model for the pore surface. The normalized volumetric water content is equal to unity at values of matric suction up to the air-entry value and equal to zero after residual saturation. The only two parameters ψ_e and D, which have obvious physical meaning, are used in Eq. (29) to express the soil-water characteristic curve (SWCC). The fractal dimension of the pore surface and the maximum pore radius can be evaluated from the pore-size distribution (PSD) measured using the mercury intrusion tests. The air-entry value can be calculated from the Young-Laplace equation using the maximum pore radius. Thus, the soil-water characteristic curve (SWCC) can be calculated from Eq. (29) using the fractal dimension and the air-entry value obtained from the mercury intrusion tests. The soil-water characteristic curves (SWCC) were calculated from Eq. (29) using the fractal dimension of the pore surface and the air-entry value for specimens of S-02, S-03 and S-04, which listed in Table 1. The value of the normalized volumetric water content is calculated from Eq. (30). Here S_r=7.5% (Watabe et al., 2000). Experimental data show in good accord with the calculation of the soil-water characteristic curve (SWCC) in Fig. 12. The experimental data of the soil-water characteristic curve (SWCC) measured in a multi-step outflow experiment were given by Vogel and Roth (1996) for silty soils. The saturated volumetric water contents are 0.502 and 0.448 for silty soils at the A and B horizons, respectively. The residual volumetric water content is equal 0 for both A and B horizons. Comparisons between the calculations ofEq. (29) and experimental data of the soil-water characteristic curves (SWCC) were shown in Fig. 13. The calculation of the soil-water characteristic curves (SWCCs) for silty soils was obtained from Eq. (29). The fractal dimensions of the soil pore and the air-entry values were listed in Table 1.

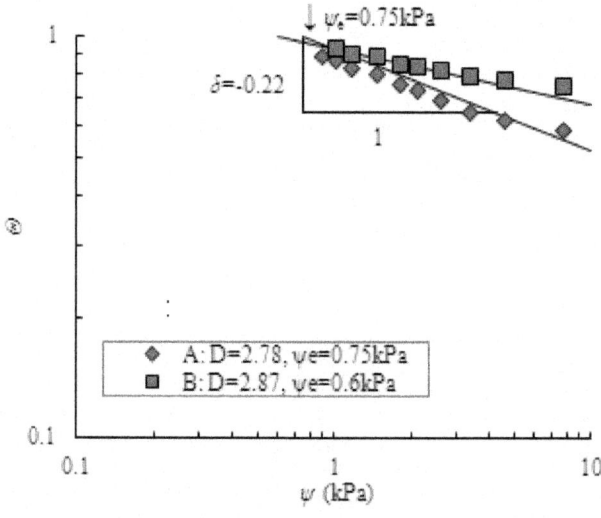

Figure 13: Comparison between prediction and experiments of SWCCs (Data from Vogel and Roth, 1996)

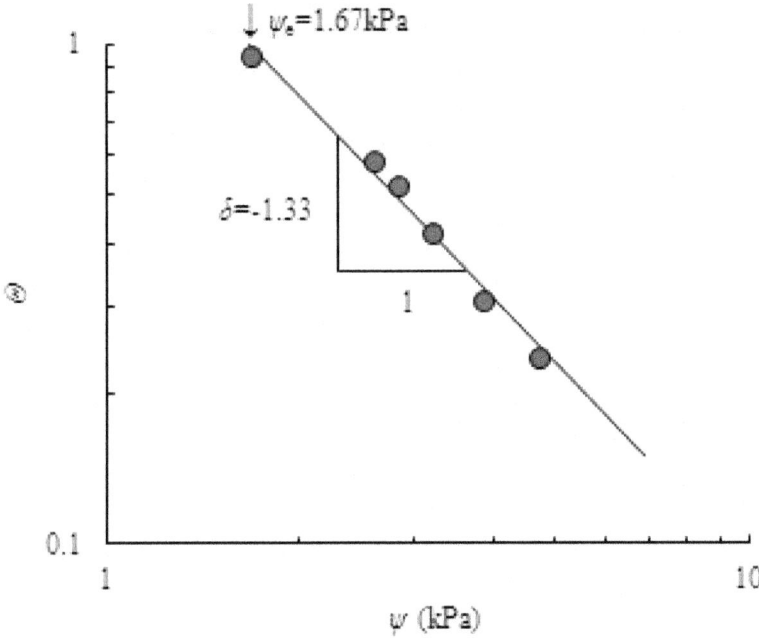

Figure 14: Comparison of the SWCCs for Toyoura sand (Data from Uno et al., 1998)

Uno et al. (1998) gave the experimental data of the soil water characteristic curves for Toyoura sand. Using the fractal dimension and the air-entry value obtained from the pore-size distribution (PSD), the soil-water characteristics can be simulated from Eq. (29) for Toyoura sand. The parameters obtained from the pore-size distribution (PSD) and used to predict unsaturated hydraulic conductivity are listed in Table 1. θ_s=0.425, θ_r=0. Comparison between the prediction of Eq. (29) and experimental data of the soil-water characteristics is shown in Fig. 14 for Toyoura sand.

Stingaciu et al. (2009) evaluated the feasibility of using nuclear magnetic resonance (NMR) relaxometry measurements to characterize pore size distribution and hydraulic properties in four porous samples with different texture and composition. The sandy samples FH31 and W3 were loaded in the penetrometer and packed at the same packing density as for the NMR measurements. The Mix8 and MZ samples were used as solid conglomerates. From the relaxation time distribution functions, the cumulative pore-size distribution functions were calculated with the average surface relaxivity. The normalized cumulative pore-size distributions functions were displayed in Fig. 15. The fractal dimension of pore surface and the maximum pore radius can be obtained from Fig. 15, and were listed in Table 2.

Table 2: Parameters for the prediction of SWCCs

Sample	Method	Bulk density (g/ cm³)	θ^s	θ^r	D	R (mm)	ψ_e (kPa)
FH31	Pressure plate	1.58	0.32	0.02	0.85	0.065	2.31
W3	MSO	1.42	0.33	0.058	1.51	0.0175	8.6
Mix8	Rosetta	1.45	0.41	0.061	1.52	0.00053	290
MZ	Pressure plate	1.60	0.44	0	2.29	0.00035	429

The soil-water characteristic curves (SWCC) were determined using the standard sand bed, pressure cell or multistep outflow method, and were plotted in Fig. 16. SWCC of FH31 and MZ were based on pressure plate measurements. SWCC of W3 was determined using multistep outflow (MSO) and that of Mix8 using ROSETTA software (Schaap et al., 2001). The matric suction as plotted in the abscissa (Fig. 16) was transformed into pore diameter using the Young-Laplace equation. In addition, the ordinate was the normalized volumetric water content. The predictions of SWCC were conducted usingEq. (29) with the parameters listed in Table 2. The parameters were derived from the pore-size distribution (PSD) shown in Fig. 15. Comparisons of the predictions and experimental data were shown in Fig. 16. The predictions of Eq. (29) were in good accord with the experimental data.

In general, the soil-water characteristic curves (SWCC) were usually interpreted as cumulative distribution Functions in comparison to pore-size distribution (PSD) functions Using the soil-water characteristic curves (SWCC), pore-size distribution can be extracted for a given porous medium on the basis of an empirical law that related the pore suction to the effective pore radius. The soil-water characteristic curves (SWCC) and the pore-size distributions (PSD) are related by correlations as: (1) the absolute values of the slopes of SWCC and PSD are equal; (2) the air-entry value of SWCC is related to the maximum pore radius of PSD through the Young-Laplace equation.

The following conclusions can be obtained from Figs. 12-16: (1) The surface of the soil pore can be described by fractal model, and the surface fractal dimension of the soil pore and the air-entry value can be evaluated from the pore-size distribution (PSD). The soil-water characteristic curves (SWCCs) can be calculated using fractal model of the soil pore. (2) The fractal dimension and air-entry value obtained from the soil pore-size distribution (PSD) are equivalent to those obtained from the soil-water characteristic curve (SWCC). Hence the fractal dimension and air-entry value can be obtained from the fitting of the soil-water characteristic curve (SWCC) if the pore-size distribution (PSD) were not measured.

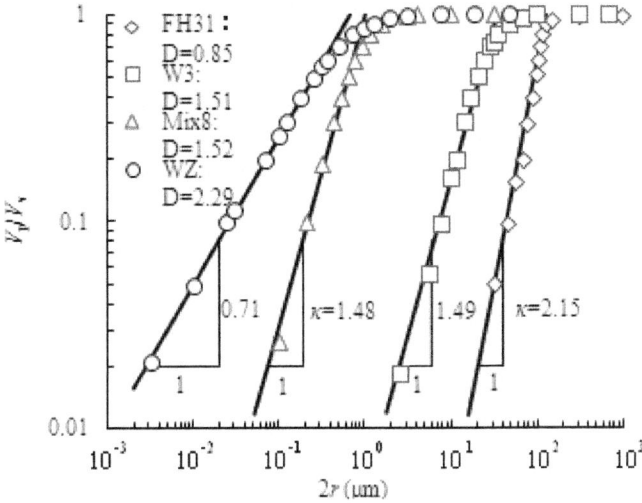

Figure 15: Cumulative pore-size distribution of for the three artificial substrates (FH31, W3, and Mix8) and the natural soil Merzenhausen (MZ), from which the fractal dimension of pore surface and the maximum pore radius were derived (Data from Stingaciu et al. 2009)

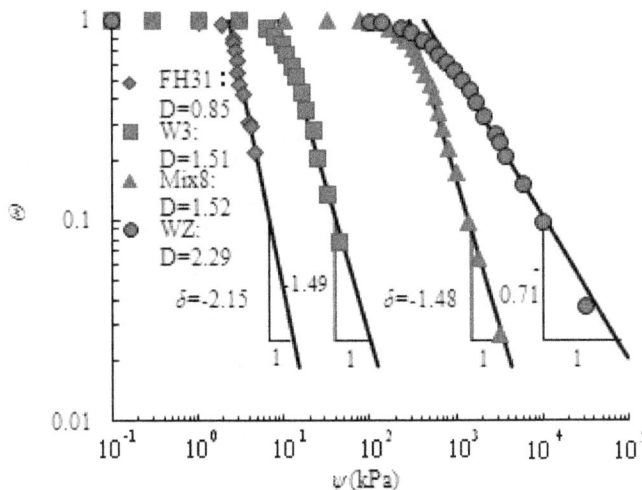

Figure 16: Comparison between the predictions of Eq. (29) and experimental data of SWCCs (Data fromStingaciu et al. 2009)

UNSATURATED HYDRAULIC CONDUCTIVITY AND DIFFUSIVITY

Brief Review

Brutsaert (2000) made a brief review of the capillary model, which was cited as follows.

It is very difficult to determine unsaturated hydraulic conductivity through experiments. Hydraulic conductivity for unsaturated soils is a function of soil-water content θ. For this reason, over the years many attempts have been made to represent the hydraulic conductivity by simple parametric equations, in terms of other properties of the soil which are easier to determine. One of these was of the following form (Brutsaert, 2000):

$$k_r = \Theta^\omega$$

(31)

In the past, the power form of Eq. (31) has been derived on the basis of some widely different conceptual models of the pore geometry. There have been some marked differences in the obtained values of ω, depending on the adopted model of the pore geometry.

Uniform pore size models

A simple approach was that porous medium was characterized by some equivalent uniform pore size and the variability of the pore sizes was not considered. This approach invariably resulted in a constant value of ω. The capillary tube model of Averyanov (Polubarinova-Kochina, 1952) led to $\omega=3.5$, whereas the extension of the hydraulic radius model of Kozeny (Kozeny, 1927) to unsaturated soils byIrmay (1954) produced $\omega=3$.

Parallel models

In parallel models, the pore system was assumed to be equivalent with a bundle of uniform capillary tubes of many different sizes. The pore-size distribution was derived from the soil-water characteristics (SWCC), which can be related to the effective pore radius r by the Yong-Laplace equation. The true mean velocity in each pore was described by a Poiseuille-like equation for creeping flow. Purcell (1949) and Gates and Tempelaar-Lietz (1950) first used this approach. Several subsequent applications of the parallel model can be written in the common form (Brutsaert, 1967):

$$k_r = \frac{\beta \int_0^{\Theta} \psi(x)^{-2} dx}{\beta_0 \int_0^1 \psi(x)^{-2} dx}$$

$$(32)$$

where x is the dummy variable representing Θ. The variable $\beta = \beta(\Theta)$ is related to the tortuosity, and β_0 is its value at satiation, when $\Theta = 1$. The tortuosity concept was originally introduced by Carman (1956) as an improvement on the non-uniform hydraulic radius model of Kozeny (1927) and it can be expressed as $T = (L_e/L)^2$, in which L_e is the actual or microscopic path length of the fluid particles in the pores and L is their apparent or macroscopic path length along the Darcy stream lines. The parameter β was introduced in the parallel model by Wyllie and Spangler (1952), because without it, Eq. (32) yielded values considerably larger than their experimental data. Burdine (1953) proposed on the basis of his experimental data that $\beta/\beta_0 = \Theta^2$.

A further development was performed by adopting the following soil-water characteristic curve (SWCC) (Brooks and Corey, 1966):

$$\Theta = 1 \quad \text{for } \psi \leq \psi_e$$

$$\Theta = \left(\frac{\psi}{\psi_e}\right)^b \quad \text{for } \psi \leq \psi_e$$

$$(33)$$

Brooks and Corey (1966) integrated (32) with Burdine's assumption for β (Burdine, 1953) to obtain Eq. (31), with an exponent

$$\omega = 3 - \frac{2}{b}$$

$$(34)$$

It should be noted that, without Burdine's assumption for the tortuosity, the parallel model Eq. (32) applied with Eq. (33) would have yielded

$$\omega = 1 - \frac{2}{b}$$

$$(35)$$

Series-parallel models

The theoretical construction of this model also started with a bundle of parallel pores, each with a different but uniform size. However, these pores were then cut normally to the direction of flow with two resulting faces, and finally after some random rearrangement of the tubes the faces are joined again. This way account was taken of the random variations of the pore sizes, not only in the plane normal to the direction of flow, but also along the direction of flow. The discharge rate in each single pore, which consisted of two sections in cases, was assumed to be governed by the section with the smaller diameter. The pore-size distribution was derived form of the soil-water characteristics by means of the Yong-Laplace equation. The true velocity in each pore was obtained by means of a Poiseuille-like equation.

This model was originally proposed by Childs and Collis-George (1950) in a finite-difference scheme to calculate hydraulic conductivity from experimental data of the soil-water characteristics (SWCC). It was subsequently reformulated in integral form by Brutsaert (1968), to allow the derivation of more concise analytical expressions for hydraulic conductivity. The integral form can be written as

$$k_w = \left[\left(\frac{2T}{\gamma}\right)^2 \frac{\theta_0^2(1-S_r)^2}{G}\right] \times \left[\int_0^\Theta \int_0^x \psi(x)^{-2}\,dy\,dx + \int_0^\Theta \psi(x)^{-2}\int_0^\Theta\,dy\,dx\right]$$

(36)

where G is a geometrical constant. In the case of Poiseuille's equation, G=8. It was trivial to show by integration by parts that the first double integral on the right was identical to the second. Hydraulic conductivity can be expressed concisely as (Brutsaert, 2000):

$$k_r = \frac{\int_0^\Theta (\Theta - x)\psi(x)^{-2}\,dx}{\int_0^1 (1-x)\psi(x)^{-2}\,dx}$$

(37)

The capillary model similar with Eq. (37) in form was originally proposed for by Fatt and Dykstra (1951).

Fractal Model for Hydraulic Conductivity

Soil pore systems are viewed as a collection of interconnected voids. The individual voids are considered to have two equilibrium states: either filled by water or empty. Flow within the soil pores is assumed to be described by (Burdine, 1953; Mualem, 1976)

$$v = -\frac{r^2 g}{c\eta}\frac{dh}{dx}$$

(38)

where v is the average velocity within the pore; r is the radius of the pore; g is the gravity acceleration; η is the kinematic viscosity; c is a constant depending on the pore geometry; h is the hydraulic head, x is the position axis. Let us consider a porous slab of thickness Δx isolated from a homogenous soil column by two parallel cross-sections normal to the x axis. The areal porosity at each face of the slab is the same and it is equal to the volumetric porosity. The major simplifying assumption is that the actual configuration of slab may be replaced by a set of capillary tubes with different radii, parallel to the xaxis. The flux dq flowing through the connections between pores of radius $r_1 \rightarrow r_1 + dr_1$ on one side of the slab (at x) and pores of radius $r_2 \rightarrow r_2 + dr_2$ on the other side of the slab (at x+dx) can be expressed by (Mualem, 1976)

$$dq = \beta r_e^2(r_1, r_2, \rho) A_e(r_1, r_2, \rho) \frac{dh}{dx} dr_1 dr_2$$

$$(39)$$

where β is a constant which incorporates the fluid with the matrix properties; $r_e(r_1, r_2, q)$ is the effective radius; $A_e(r_1, r_2, q)$ is the effective area; ρ is the maximum radius of the water filled pores at volumetric water content θ. According to the Darcy equation, the hydraulic conductivity k_w is given by

$$k_w = \frac{q}{dh/dx} = \beta \int_0^\rho \int_0^\rho r_e^2(r_1, r_2, \rho) A_e(r_1, r_2, \rho) dr_1 dr_2$$

$$(40)$$

The relatively hydraulic conductivity (kr) is the ratio of the hydraulic conductivity at any volumetric water content (kw) to the hydraulic conductivity at saturation (ks). The relative hydraulic conductivity (RHC) is related to the radius of the soil pores, and is written as (Mualem, 1976)

$$k_r = \frac{k_w}{k_s} = \frac{\int_0^r \int_0^r r_e^2(r_1, r_2, r) A_e(r_1, r_2, r) dr_1 dr_2}{\int_0^R \int_0^R r_e^2(r_1, r_2, r) A_e(r_1, r_2, r) dr_1 dr_2}$$

$$(41)$$

where r is the maximum radius of the water filled pores at the volumetric water content θ. It is assumed that the pore distribution at the two cross-section is completely random. The probability of pores of radius $r_1 \rightarrow r_1 + dr_1$ at x to encounter pores of radius $r_2 \rightarrow r_2 + dr_2$ at x+dx is given by

$$A_e(r_1, r_2) dr_1 dr_2 = f(r_1) f(r_2) dr_1 dr_2$$

$$(42)$$

The effective radius is assumed to be equal to the radius r reduced by a factor, which is equal to the ratio between the effective area to flow and the actual pore area. The effective radius is given by (Mualem, 1976)

$$r_e = r \frac{\int_0^r \int_0^r A_e(r_1, r_2, r) dr_1 dr_2}{\int_0^R f(r_1) dr_1} = r \Lambda^{1/2}$$

$$(43)$$

Substituting Young-Laplace equation and Eqs.(42) and (43) into Eq.(41), the relative hydraulic conductivity (RHC) is obtained as

$$k_r = \left(\frac{\psi}{\psi_e}\right)^{\xi}$$

(44)

where $\xi=3D-11$. Substituting Eq. (29) into Eq. (44), the relative hydraulic conductivity (RHC) unction expressed by effective degree of saturation is written as follows

$$k_r = \Theta^{\xi/\delta}$$

(45)

Equations (44) and (45) are the relative hydraulic conductivity (RHC) function derived from the fractal model of the pore surface. The parameters in the proposed hydraulic conductivity function have an obvious physical implication, and can be determined from the pore-size distribution (PSD). Eqs. (44)and (45) offer a powerful, physical-based method to calculate the relative hydraulic conductivity (RHC). Eq. (44) is nearly similar with the Brooks and Corey equation (Brooks and Corey, 1966) in form, and the parameters in Eq. (44) have obvious physical implication. Eqs. (44) and (45) offer a powerful, physical-based method to calculate the relative hydraulic conductivity (RHC).

The expression of the soil-water diffusivity (SWD) can be derived from the unsaturated hydraulic conductivity and soil-water characteristics, and written as follows:

$$d = k_w \left|\frac{d\psi}{d\theta}\right|$$

(46)

where d is the soil-water diffusivity (SWD), k_w is the saturated hydraulic conductivity. SubstitutingEqs. (44) and (45) into Eq. (46), the soil-water diffusivity (SWD) is written as

$$d = k_s \psi_e \left(\frac{\psi}{\psi_e}\right)^{\varsigma}$$

(47)

$$d = k_s \psi_e \Theta^{\varsigma/\delta}$$

(48)

where $\varsigma=2D-7$.

As given in Eqs. (29), (44), (45), (47) and (48), methods to evaluate the soil-water characteristics (SWCCs), unsaturated hydraulic conductivity (RHC) and soil-water diffusivity (SWD) are proposed using the fractal model for the pore surface. The fractal dimension and the maximum radius of soil pores can be obtained from the pore-size distribution (PSD). The air-entry value can be calculated from the Young-Laplace equation using the maximum radius of soil pores. The soil-water characteristic curves (SWCC), unsaturated hydraulic conductivity (RHC) and soil-water diffusivity (SWD) can be predicted using the fractal dimension and air-entry value, which can be determined from the pore-size distribution (PSD). According to Eq. (29), the fractal dimension of the pore surface and the air-entry value can also be evaluated from the fitted soil-water characteristic curve (SWCC). Thus, the unsaturated hydraulic conductivity (RHC) and soil-water diffusivity (SWD) can also be estimated fromEqs. (44), (45), (47) and (48) using the fractal dimension and air-entry value evaluated from the soil-water characteristic curve (SWCC).

Predictions of SWCC and RHC from PSD

Vogel and Roth (1998) inferred effective hydraulic properties of unsaturated soil from the structure of the pore surface. The water retention characteristic and the hydraulic conductivity were simulated by network models with $32^3 = 32768$ nodes. The hydraulic conductivity, k_w, is determined at each step of desaturation by imposing a pressure gradient Δp across the ends of the network. Water flow q_{ij} in a bond between the nodes i and j through a horizontal surface A is described by Poiscuille's law. The hydraulic conductivity was calculated as $= qL/(A\Delta p)$, where L is the vertical length of the network. Comparison between the predictions of Eq. (44) and experimental data given by Vogel and Roth (1998) was shown in Fig. 17. The parameters used to do prediction were listed in Table 1, which were obtained from the fractal model for the pore-size distribution (PSD). The predictions were in good accord with the experimental data.

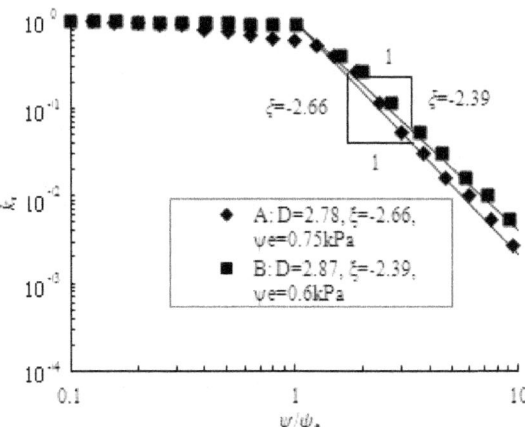

Figure 17: Comparison between the predictions of Eq. (44) and experimental data of the RHC (Data from Vogel and Roth, 1998)

Uno et al. (1998) studied the relationship between the pore-size distribution (PSD) and unsaturated hydraulic properties of Toyoura sand. The relationship between the relative hydraulic conductivity (RHC) and the normollized volumetric water content Θ can be calculated from Eq. (45) using the parameters listed in Table 1. The prediction of Eq. (45) satisfactorily agrees with the experimental data of relative hydraulic conductivity (RHC) for Toyoura sand in Fig. 18. The proposed function of relative hydraulic conductivity (RHC) (Eq. (45)) is verified by the experimental data in Fig. 18.

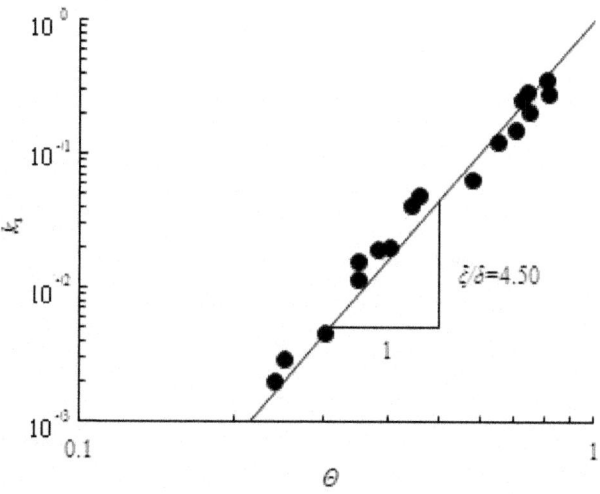

Figure 18: Comparison between predictions of Eq. (45) and experimental data of the RHC for Toyoura sand. (Data from Uno et al., 1998)

Prediction of RHC and SWD from SWCC

The characteristics of the pore-size distribution (PSD) were prerequisite to determine the relative hydraulic conductivity and soil-water diffusivity of unsaturated soils in Eqs. (44), (45), (47) and (48). A close relationship was established between the soil-water characteristic curve (SWCC) and the pore-size distribution by fractal model for the pore surface. The pore-size distribution (PSD) can be estimated from the soil-water characteristic curve because many soils show that the soil-water characteristic curve is equivalent to the pore-size distribution (PSD) function (Watabe et al., 2000). Thus, the fractal dimension of the pore surface and the air-entry value can be estimated from the fitting of the soil-water characteristic curve (SWCC). Thus, unsaturated hydraulic properties, such as relative hydraulic conductivity (RHC) and soil-water diffusivity (SWD) can be predicted using the fractal dimension and air-entry value obtained from the soil-water characteristic curve (SWCC).

Smettem and Kirkby (1990) offered the hydraulic properties of haploxeroll loam. Experimental data of the soil-water characteristic curve (SWCC) were shown in Fig. 19. The fractal dimension of the pore surface can be obtained from the fitted soil-water characteristic curve (SWCC). Through fitting the soil-water characteristic curve (SWCC), we obtained that δ=-0.37 for haploxeroll loam from Fig. 19. Fractal dimension of the pore surface can be calculated from Eq. (29) using the fitting value of δ. The fractal dimension (D) of the pore surface is 2.63 for haploxeroll loam. The air-entry value is 0.14kPa obtained from Fig. 19.

The hydraulic properties of unsaturated soils can be predicted using the fractal dimension of the pore-size distribution (PSD), obtained from fitted soil-water characteristic curve (SWCC). The fundamental parameters for predictions of hydraulic properties were tabulated in Table 3. The comparisons between predicted result and experimental data of the relative hydraulic conductivity (RHC) were shown in Fig. 20. The coefficient of hydraulic conductivity at saturation is 140 mmh^{-1}. Fig. 20 depicted the relationship between the relative hydraulic conductivity (RHC) and water potential for haploxeroll loam. The parameter ξ was -3.11 calculated from Eq. (44) using the fractal dimension D=2.63 for haploxeroll loam. The predictions of the relative hydraulic conductivity (RHC) calculated from the proposed function Eq. (44) are in satisfactory agreement with the experimental data.

The relationship between the soil-water diffusivity and water potential for haploxeroll loam was also shown in Fig. 20. The values of k_s and ψ_e are 140mmh^{-1} and 0.14kPa, respectively, used in Eq. (47). The parameters ς was -1.74, calculated from Eq. (47) using the fractal dimension D=2.63. The prediction of the soil-water diffusivity was the line passing through the point of (0.14, 0.64) with the slope of -1.74. The prediction of the soil-water diffusivity nearly agreed with the experimental data. The proposed function for the soil-water diffusivity was verified by the comparison between prediction and measured results. The parameters obtained from the soil-water characteristic curve (SWCC) and used to predict the relative hydraulic conductivity (RHC) and soil-water diffusivity (SWD) for haploxeroll loam were listed in Table 3.

Table 3: Parameters used to predict the unsaturated hydraulic conductivity

Soil type	D	δ	ξ	ξ/δ	ς	$\psi_e(kPa)$
Haploxeroll loam	2.63	-0.37	-3.11	8.41	-1.74	0.14
Toyoura sand (wetting)	1.60	-1.40	-6.2	4.43	-3.8	/
Toyoura sand (drying)	1.26	-1.74	-7.22	4.15	-4.48	/

Loamy sand	2.68	-0.32	-2.96	9.25	-1.64	2.0
McGee Ranch soil	2.51(GSD)	-0.49	-3.47	7.08	-1.98	4.6

Figure 21 shows the experimental data and the fitting of the soil-water characteristic curve (SWCC) for loamy sand. The experimental data are scaled from Simunek et al. (1999). The parameter δ was -0.32, which is obtained from the fitting of the soil-water characteristic curve (SWCC) for loamy sand fromFig. 26. Hence the fractal dimension of the pore surface was 2.68 calculated from Eq. (29) using the fitting value of δ for loamy sand. The air-entry value was given by the intersection between the line expressed by Eq. (29) and the line of $\Theta=1$ in log-log plot. The air-entry value is 2kPa obtained from Fig. 21.

The parameters used to predict the hydraulic properties of loamy sand were listed in Table 3. Substituting D=2.68 and ψ_e=2kPa in Eq. (44), the predictions of relative hydraulic conductivity (RHC) can be obtained. The comparisons between predictions and experimental data of the relative hydraulic conductivity were shown in Fig. 22. The parameter ξ is -2.96 calculated from Eq. (44) using the fractal dimension D=2.68 for loamy sand. The predictions of the relative hydraulic conductivity were in good accord with the experimental data. The proposed function for the relative hydraulic conductivity was validated by the good agreement between the predictions and measurements.

The parameters of fractal model were first calibrated by matching the measured moisture-suction data. It should be noted that the experimental values of hydraulic conductivity were not matched by adjusting the model parameters. The calibrated SWCC model parameters were, instead, directly used to predict the relative hydraulic conductivity. The soil-water characteristic curves (SWCC) and their calibration of eight soils were shown in Fig. 23. The calibration of SWCC model parameters were listed in Table 4. As seen in these figures, the measured moisture-suction data for these soils were unavailable for the full range (0%-100%) of the degree of saturation. Predicting the suction beyond the available experimental data range was a challenging task since the pattern of variation was unknown (Ravichandran and Krishnapillai, 2011).

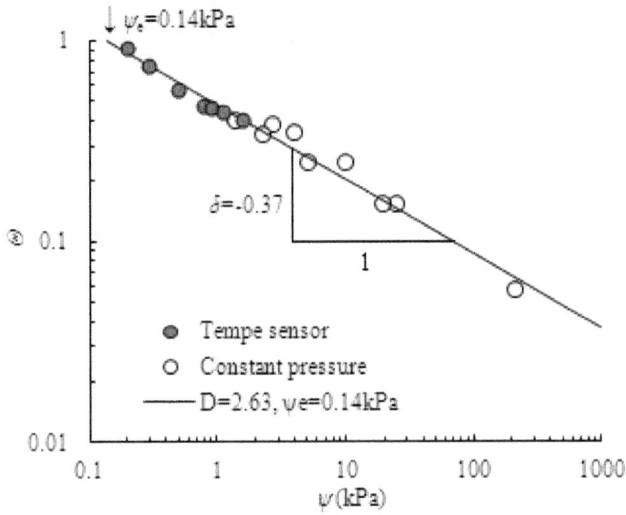

Figure 19: Fractal dimension of the pore surface and air-entry value evaluated from SWCC (Data from Smettem and Kirkby, 1990)

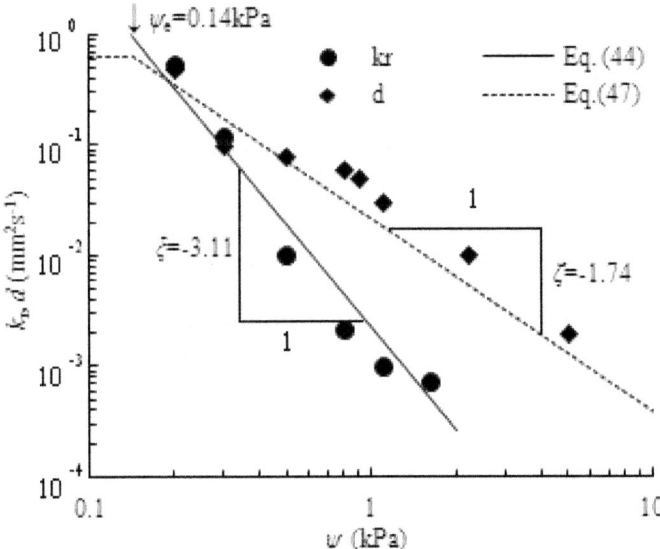

Figure 20: between predictions and experimental data of RHC and SWD for haploxe-roll loam (Smettem and Kirkby, 1990)

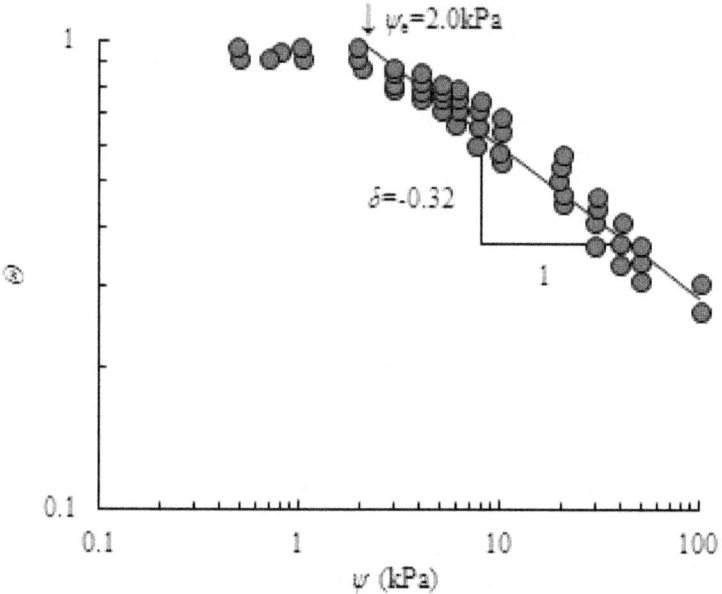

Figure 21: Fractal dimension and the air-entry value obtained from the fitting of SWCC for of loamy sand (Data from Simunek et al., 1999)

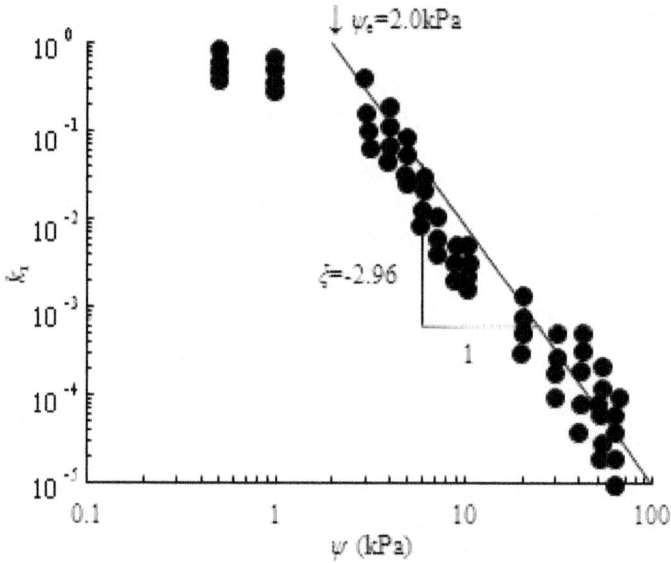

Figure 22: Comparison between measurements of RHC and prediction of Eq. (44) for loamy sand (Data fromSimunek et al., 1999)

Table 4: Parameters of the fractal model for eight selected soils

Soil	$_\theta S$	$S_r(\%)$	D	$\psi_e(kPa)$	Reference
Lakeland	0.375	30-100	1.75	2.5	Elzeftawy and Cartwright, 1981
Superstition	0.5	30-100	2.7	1.2	Richards, 1952
Columbia sandy loam	0.458	50-100	2.29	4.5	Brooks and Corey, 1964
Touchet silt loam	0.43	20-100	2.09	7	Brooks and Corey, 1964
Silt loam	0.396	50-100	2.59	15	Reisenauer, 1963
Guelph loam	0.52	45-100	2.75	3	Elrick and Bow-mann, 1964
Yolo light clay	0.375	45-100	2.87	1.5	Moore, 1939
Speswhite kaolin	0.56	55-100	2.85	15	Peroni et al., 2003

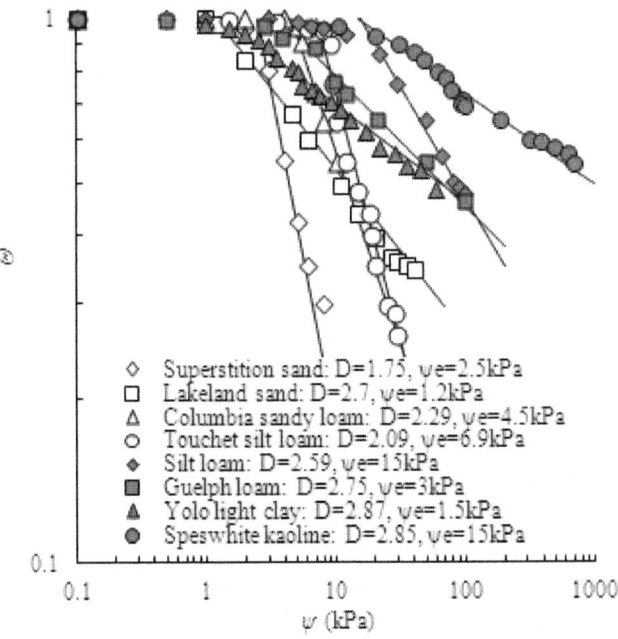

◇ Superstition sand: D=1.75, ψe=2.5kPa
□ Lakeland sand: D=2.7, ψe=1.2kPa
△ Columbia sandy loam: D=2.29, ψe=4.5kPa
○ Touchet silt loam: D=2.09, ψe=6.9kPa
◆ Silt loam: D=2.59, ψe=15kPa
■ Guelph loam: D=2.75, ψe=3kPa
▲ Yolo light clay: D=2.87, ψe=1.5kPa
● Speswhite kaoline: D=2.85, ψe=15kPa

Figure 23: Fractal dimension and air-entry value obtained from the fitting of SWCC

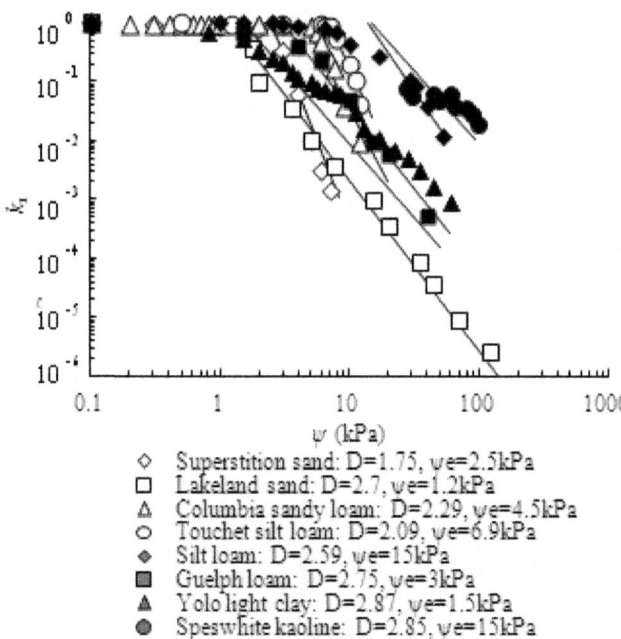

◇   Superstition sand: D=1.75, ψe=2.5kPa
□   Lakeland sand: D=2.7, ψe=1.2kPa
△   Columbia sandy loam: D=2.29, ψe=4.5kPa
○   Touchet silt loam: D=2.09, ψe=6.9kPa
◆   Silt loam: D=2.59, ψe=15kPa
■   Guelph loam: D=2.75, ψe=3kPa
▲   Yolo light clay: D=2.87, ψe=1.5kPa
●   Speswhite kaoline: D=2.85, ψe=15kPa

Figure 24: Comparison between predictions of Eq. (44) experiments of RHC

The relative hydraulic conductivity of the above mentioned eight soils were predicted using the fractal model with the same fitting parameters listed in Table 4. The fitting parameters were obtained by matching the experimental SWCC. Fig. 24 illustrated the prediction of Eq. (44) and experimental data of relative hydraulic conductivity. It should be noted that the fractal model parameters were not calibrated or adjusted to match the measured hydraulic conductivity values. The predictions of Eq. (44) were comparable for sand and loam, as shown in Fig. 24. The fractal model showed slight deviations from the measured data of clay. In the case of Yolo light clay and Speswhite kaolin, the difference between the predictions and the experimental data increased as the suction increases (Fig. 24). The fractal model showed better predictions at lower suction range. However, the accuracy of fractal model in the higher suction range could not be verified because the experimental results were available only for the lower suction ranges.

It is seen that it is feasible to estimate the relative hydraulic conductivity of unsaturated soils using the fractal dimension of the soil pore surface from Figs. 19-24. A practical method to express the relative hydraulic conductivity was proposed to use the fractal dimension and the air-entry value, which can be obtained from the fitting of the soil-water characteristic curve (SWCC).

Prediction of RHC and SWD from GSD

The feasibility to determine the soil hydraulic properties using the fractal dimension of the grain-size distribution (GSD) is examined by the experimental data of McGee Ranch soil. The grain-size distribution (GSD) was measured by Hunt and Gee (2002) and was shown in Fig. 25. The fractal dimension of the grain-size distribution (GSD) is 2.51, obtained from Fig. 25.

Hunt and Gee (2002) gave the values of θ_s and θ_r were 0.4 and 0.01, respectively, the air-entry value was 4.6kPa. Using the fractal dimension of the grain-size distribution (GSD), the soil-water characteristic curve (SWCC) and the relative hydraulic conductivity (RHC) can be determined. The parameters of the grain-size distribution (GSD) used to determine the hydraulic properties are listed in Table 3. Comparisons between the predictions of Eq. (29) using the fractal dimension of the grain-size distribution (GSD) and the experimental data of soil-water characteristic curve (SWCC) are shown in Fig. 26. It was seen that a good result is obtained to predict the soil-water characteristic curve (SWCC) using the fractal dimension of the grain-size distribution (GSD).

Hunt and Gee (2002) gave the values of saturated hydraulic conductivity was 0.001cm/s. Comparisons between the predictions of Eqs. (44) and (45) and experimental data were shown in Fig. 27. The prediction of Eqs. (44) and (45) nearly satisfied the experimental data of relative hydraulic conductivity in Fig. 27. In Fig.27, the predictions of Eq. (45) deviated from the experimental data near saturation, and extend to equal to the experimental results at low water content. The soil hydraulic conductivity can be approximately determined using the fractal dimension of the grain-size distribution (GSD).

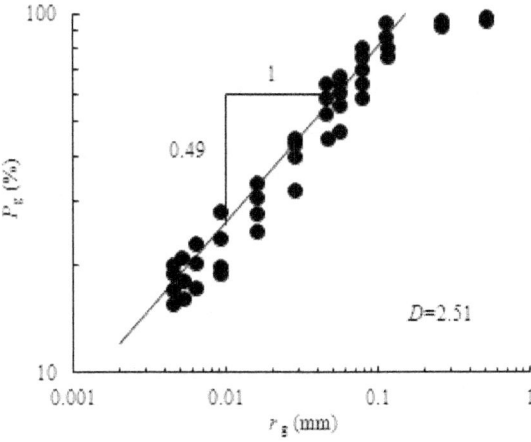

Figure 25: Fractal dimension of GSD for McGee Ranch soil (Data from Hunt and Gee, 2002)

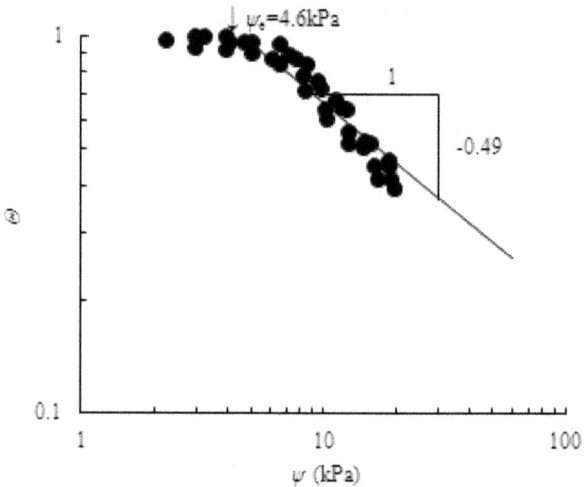

Figure 26: Comparisons between predictions and experiments of SWCC for McGee Ranch soil (Data from Hunt and Gee, 2002)

COMPARISON OF THE FRACTAL MODEL WITH THE VAN GENUCHTEN–MUALEM (G-M) MODEL

Van Genuchten–Mualem (G-M) Model for Relative Hydraulic Conductivity (RHC)

Van Genuchten (1980) derived an empirical relationship to describe the soil-water characteristic curve (SWCC):

$$\Theta = \frac{1}{(1+|\alpha\psi|^n)^m}$$

(49)

where α, n and m are the van Genuchten curve-fitting parameters, and m=1-1/n.

Burdine (1953) and Mualem (1976) presented a similar model to estimate the unsaturated hydraulic conductivity from pore size distributions inferred from soil water retention characteristics. The models share a number of similarities, allowing them to be written in a general form as (Hoffmann-Riem et al., 1999)

$$k_r = \frac{k_w}{k_s} = \Theta^l \left(\frac{\int_0^\Theta \psi^{-\chi} d\Theta}{\int_0^1 \psi^{-\chi} d\Theta} \right)^\gamma$$

(50)

where k_w is the unsaturated hydraulic conductivity at any water content, k_s is the saturated hydraulic conductivity, Θ is the normalized water content, ψ is matric suction. Generally accepted parameter values for the parameters (l, χ, γ) are $(0.5, 1, 2)$ and $(2, 2, 1)$ for the Mualem (1976) and the Burdine (1953) models, respectively.

Using the Mualem model, van Genuchten (1980) derived a closed-form to determine the relative hydraulic conductivity (RHC) at a degree of saturation,

$$k_r = \frac{\left[1 - |\alpha\psi|^{n-1}\left(1 + |\alpha\psi|^n\right)^{-m}\right]^2}{\left(1 + |\alpha\psi|^n\right)^m} \tag{51}$$

$$k_r = \Theta^{\frac{1}{2}}\left[1 - \left(1 - \Theta^{\frac{1}{m}}\right)^m\right]^2 \tag{52}$$

Equations (51) and (52) are the representations of the G-M model, which is the most widely and popularly used to predict the hydraulic conductivity.

The other models for relative hydraulic conductivity (RHC) were also introduced as follows.

Van Genuchten–Burdine (G-B) Model for Relative Hydraulic Conductivity (RHC)

Hydraulic conductivity resulting from the Burdine capillary models (Burdine, 1953) was expressed as

$$k_r = \Theta^2\left[1 - \left(1 - \Theta^{\frac{1}{m}}\right)^m\right] \tag{53}$$

Replacing Eq. (49) in Eq. (53) yielded

$$k_r = \frac{\left[1 - |\alpha\psi|^{n-2}\left(1 + |\alpha\psi|^n\right)^{-m}\right]}{\left(1 + |\alpha\psi|^n\right)^{2m}} \tag{54}$$

where $m = 1 - 2/n$.

Note that for both Eqs. (52) and (54), because m is less than 1, the derivative $dk/d\theta$ is infinite for θ_s, whatever the value of n.

Van Genuchten–Fatt & Dykstra (G-Fd) Model for Relative Hydraulic Conductivity (RHC)

The capillary model proposed by Fatt and Dykstra (1951) was expressed as

$$k_r = \frac{\int_0^{\Theta} (\Theta - x)\psi(x)^{-2-c} dx}{\int_0^1 (1-x)\psi(x)^{-2-c} dx} \tag{55}$$

Evaluating $\psi(x)$ from Eq. (49), the integrals of Eq. (55) with the condition of m>-(2+c)/2n and n>2+cresulted in:

$$k_r = \frac{\Theta B(a_1, b) I_{\Theta^{1/m}}(a_1, b) - B(a_2, b) I_{\Theta^{1/m}}(a_2, b)}{B(a_1, b) - B(a_2, b)} \tag{56}$$

where a_1=m+(2+c)/n, a_2=2m+(2+c)/n, b=1-(2+c)/n, $I_x(a, b)$ and $B(a, b)$ are the Incomplete Beta and Beta functions of positive arguments a and b, respectively, and given by:

$$B(a, b) = \int_0^1 x^{a-1}(1-x)^{b-1} dx, \quad I_y(a, b) = \frac{\int_0^y x^{a-1}(1-x)^{b-1} dx}{B(a, b)} \tag{57}$$

The results presented by Touma (2009) were obtained with c=0.5 and a_1=1.

Even though the conductivity predicted by the capillary model of Fatt and Dykstra (1951) gave the best results compared with both the quasi-analytical solution and observations, the resulting expression was not easy to use because it was necessary to evaluate Incomplete Beta and Beta functions. In order to simplify G-FD model, Θ was expressed by Eq. (49) and k_r was according to Eq. (31) with ω=2+2.5/mn. The fitted result in a value of b was close to -mn, and a conductivity curve close to that predicted by Eq. (56).

Brooks & Corey-Brutsaert Model for Relative Hydraulic Conductivity (RHC)

Brutsaert (2000) presented a capillary model expressed as Eq. (37). Combining this capillary model with the soil-water characteristic curve (SWCC) of BC model (Brooks and Corey, 1966), relative hydraulic conductivity was written as Eq. (31) in form, and the parameter ω was given by

$$\omega = 2 - \frac{2}{b} \tag{58}$$

Combining of BC model for the soil-water characteristics with the Mualem and the Burdine capillary models, the relative hydraulic conductivity curves were also expressed as Eq. (31) in form, and the parameter ω was given by

$$\omega = 3 + \frac{2}{mm} \qquad (59)$$

when BC is applied with the Burdine condition, and

$$\omega = 2.5 + \frac{2}{mm} \qquad (60)$$

when applied with the Mualem condition.

Here we focused on the G-M model, as it was the most widely used in soil science and hydrology. van Genuchten (1980) found that the predictions of the G–M model were less accurate for Beit Netofa clay. The fractal dimension and the air-entry value evaluated from the fitted soil-water characteristic curve (SWCC) are 2.89 and 60kPa, respectively (Fig. 28) The fitted parameters of the van Genuchten model, α and n are 0.0015 and 1.17, respectively from Fig.28. The parameters for the determination of the relative hydraulic conductivity (RHC) are listed in Table 5.

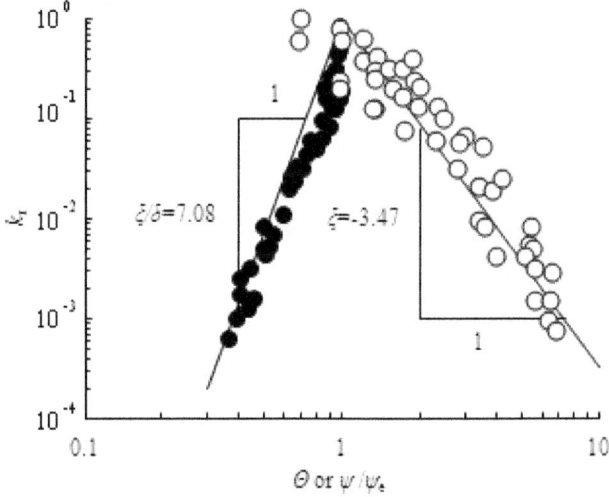

Figure 27: Comparisons between predictions and experimental data of RHC for McGee Ranch soil (Data fromHunt and Gee, 2002)

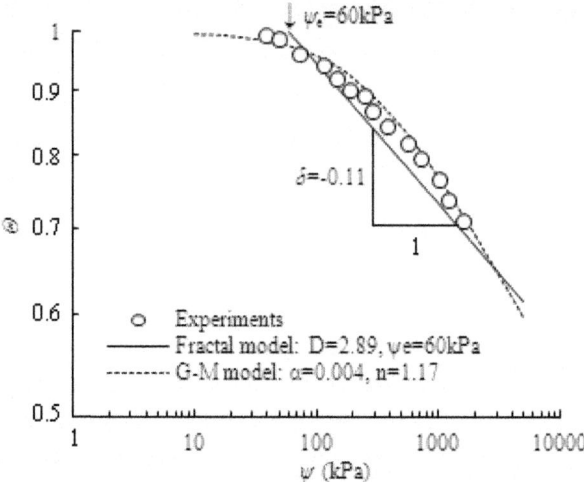

Figure 28: Fractal dimension and the parameters of G-M model obtained from SWCC for Beit Netofa clay (Data from van Genuchten, 1980)

Table 5: Parameters obatined from SWCC for prediction of RHC

Soil type	Fractal model				G-M model	
	δ	D	ξ	ξ/δ	α (kPa⁻¹)	n
Beit Netofa clay	-0.11	2.89	-2.33	21.2	0.004	1.17
Hanford glacial soil 0-099	-0.62	2.38	-3.86	6.23	0.015	1.71
Hanford glacial soil 2-1637	-0.82	2.18	-4.46	5.44	0.076	1.89
Grenoble sand	-0.64	2.36	-3.92	6.13	0.4	2.17

The saturated and residual water contents of Beit Netofa clay are 0.446 and 0, respectively. Comparisons between the predictions of both the fractal model (Eq. (44)) and the G–M model (Eq. (51)) and the experimental data of the relative hydraulic conductivity (RHC) for Beit Netofa clay are shown in Fig. 29. It is seen that predictions of the fractal model (Eq. (44)) satisfactorily agree with the experimental data, especially at high matric suction, while the predictions of the G–M model (Eq. (51)) are found to deviate from the experimental data. However, the fractal model shows a sharp corner at the air-entry value point, which is disagree with the measured data that usually show smooth transition after the air-entry value point. This default does not reduce the advantages to predict the relative hydraulic conductivity (RHC) using the fractal dimension obtained from the pore-size distribution (PSD), because the hydraulic conductivity near the point of the air-entry value nearly equal to the saturated conductivity.

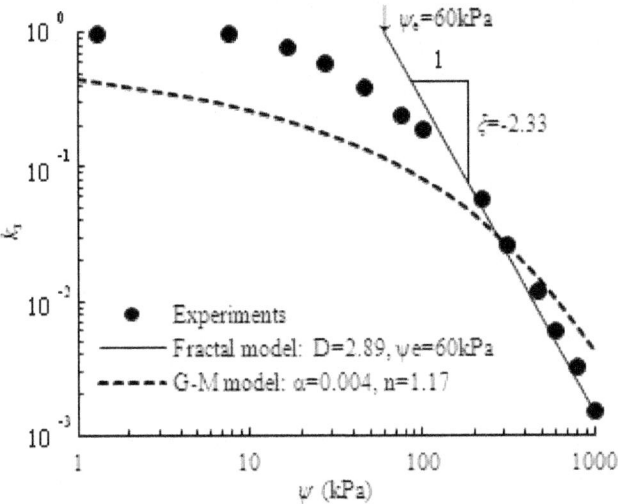

Figure 29: Comparison between the predictions of the RHC using fractal model and G-M model (van Genuchten, 1980)

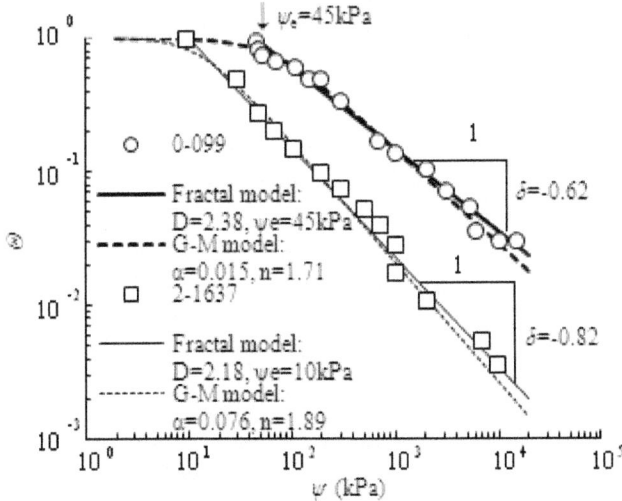

Figure 30: Fractal dimension and G-M parameters obtained from SWCC for Hanford glacial (Khaleel and Relyea, 1995)

The measured soil–water characteristic curves of Hanford glacial soil by Khaleel and Relyea (1995) were shown in Fig. 30. The saturated water contents (θ_s) are 0.338 and 0.303 for the samples 0–099 and 2–1637, respectively. The residual water contents (θ_r) are 0.039 and 0.025 for the samples 0–099 and 2–1637, respectively. From Fig. 30, it is seen that the soil-water characteristic curves (SWCC) of Hanford glacial soil are in good accord with the fractal model. The fractal dimensions (D), according to Eq. (29), are 2.35 and 2.18 obtained from the soil-water characteristic curves for the samples 0–099 and 2–1637 of Hanford glacial soil, respectively.

From the experimental data of the soil-water characteristic curve (SWCC), the parameters for determining relative hydraulic conductivity (RHC) using the fractal dimension are listed in Table 5. The larger the fractal dimension, the larger the complexity of the pore connectivity will be. The larger the fractal dimension, the larger the change in the water content with the same change in matric suction will occur. From the fittings of the soil–water characteristic curve in Fig. 30, the parameters (a and n) in the van Genuchten model (1980) for the soil–water characteristic curve (SWCC) are obtained, and are also listed in Table 5. Comparisons between the predictions of the fractal model (Eq. (45)) and the experimental data of the relative hydraulic conductivity (RHC) for Hanford glacial soil are shown inFig. 31. The predictions of the fractal model (Eq. (45)) nearly agree with the experimental data of the relative hydraulic conductivity (RHC) for Hanford glacial soil in Fig. 31. The predictions of the G–M model (Eq. (52)) deviate from the experimental data of the relative hydraulic conductivity (RHC) inFig. 31. It is seen that the predictions of the fractal model are better than those of the G–M model for Hanford glacial soil in Fig. 31.

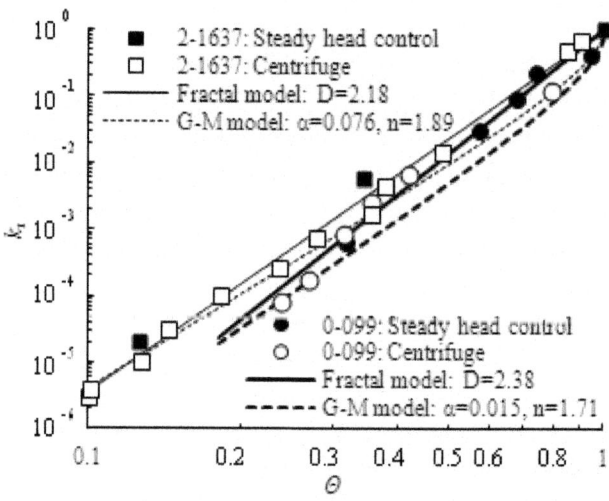

Figure 31: Comparison between the predictions using fractal model and G-M model (Data from Khaleel and Relyea, 1995)

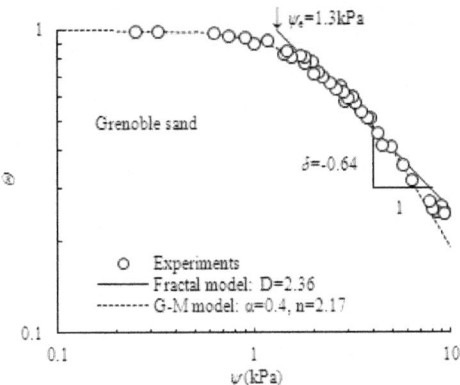

Figure 32: Fitted SWCC curves

The combinations of Eq. (29) with fractal capillary model were tested on Grenoble sand. Data points for the soil-water characteristic curve (SWCC) and hydraulic conductivity were taken from Touma (2009). The infiltration experiment was conducted under a constant head of 2.3cm (Touma & Vauclin, 1986). The solid lines in Figure 37 were the fitted curves for Eq. (29), and the resulting predicted hydraulic conductivity were shown in Fig. 32. Fractal model gave good results for the Grenoble sand. The parameters obtained through the fitting results of soil-water characteristic curves (SWCC) were used to predict hydraulic conductivity curve. Comparison between the fractal model and the G-M model were shown in Figs. 33. It was found that the prediction of hydraulic conductivity curve using Eq. (45) closed to the experiments for Grenoble sand.

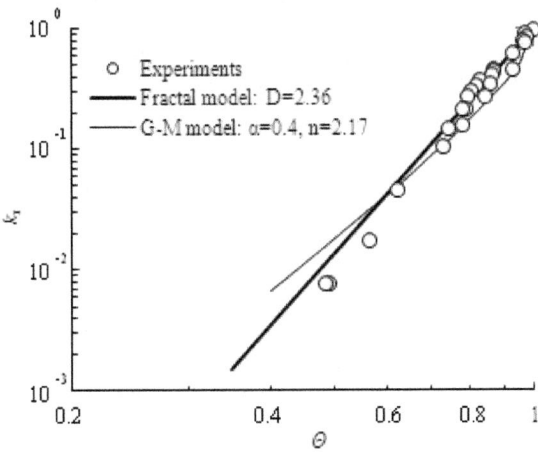

Figure 33: Comparison between fractal model and G-M model

RAINFALL INFILTRATION OF UNSATURATED SOILS

Rainfall induced slope failures often occur as relatively shallow failure surfaces orientated parallel to the slope surface and are observed and analyzed by different mechanisms. The effect of seepage on slope stability is introduced in the analyses by calculating the critical depth for an infinite slope with rainfall infiltration, while the effect of both negative and positive pore water pressures on the stabilities of initially unsaturated slopes are explained and coupled with infinite slope analysis and pore-air flow analysis methods in order to present a predictive formulation of slope failures that occurs in rainfall evens and is derived from a fractal model on unsaturated soil. The formulation serves as a baseline analysis method for evaluating potentially unstable slopes.

The pore water pressure pattern that develops in the initially unsaturated soil will occur as a transient process as the infiltration moves downward into the soil profile. Several factors should been taken into account in unsaturated analyses. The shear strength of the soil mass and the development of seepage forces which both depend on the evolution of the pore water pressure profile must been addressed in detail.

Here, an individual soil slice can be treated as a one-dimensional column with vertical infiltration. Suppose that any lateral flow between the adjacent slices will be equal on the up-slope and down-slope boundaries. Then considering vertical infiltration only could meet the flow continuity requirement. These one-dimensional infiltration analyses are addressed by the saturated/unsaturated seepage finite element software SEEP/W (GEO-SLOPE International Ltd.2007) coupled with the air flow software AIR/W (GEO-SLOPE International Ltd.2007).

For a homogenous, isotropic soil, one-dimensional water flow equation (GEO-SLOPE International Ltd.2007) is given by:

$$m_w \gamma_w \frac{\partial H_w}{\partial t} = \frac{\partial}{\partial y}\left(k_w \frac{\partial H_w}{\partial t} \right) + m_w \frac{\partial P_a}{\partial t} + Q_w$$

(61)

where H_w is the total water energy potential comprised of both pressure and elevation potentials; P_a is the pore air pressure; kw is hydraulic conductivity; y is y coordinate; m_w is the slope of SWCC; t is time; and Q_w has units of length per time.

For the air conversation of mass, we can arrive at the pore air general mass balance equation (GEO-SLOPE International Ltd.2007) as like:

$$\left(\rho_w m_w + \frac{\theta_a}{\bar{R}T}\right)\frac{\partial P_a}{\partial t} = \frac{\partial}{\partial y}\left[\frac{\rho_a k_a}{\gamma_{0a}}\frac{\partial P_a}{\partial y} + \frac{\rho_a^2 k_a}{\rho_{0a}}\right] + \rho_a \lambda_w m_w \frac{\partial H_w}{\partial t}$$

(62)

where k_a is air permeability; θ_a is volumetric air content; ρ_a is air density; γ_{oa} is initial air unit weight; ρ_{oa} is initial air density; $\bar{R}$ is ideal air constant and $\bar{R}$ =287J/(kg K) for dry air; T is temperature; H_w is total head and P_a is pore air pressure. These two governing equations can be used to determine total pressure head profile for the water and the pore air pressure, and then suction is attained.

The relationships relating volumetric water content and hydraulic conductivity to suction must be known to solve the Eq. (61). And similarly to calculate the pore air pressure the pore air permeability function is also wanted. To find the effect of different fractal dimensions and air-entry values of different soil types, series of parameter studies have conducted. Different types of Soil-water characteristic curves and hydraulic conductivity curves were shown in Figs. 34 and 35. In the legends of these pictures, Soil 10, 2.1,-5, for example, means a type of soil with air-entry value, ψ_e=10kPa, the fractal dimension, D=2.1 and saturated hydraulic conductivity k_s=1×10⁻⁵m/s. It is assumed that θ_r=0. The pore air permeability function curve used in these studies was shown in Fig.36, and it remained a constant (GEO-SLOPE International Ltd.2007).

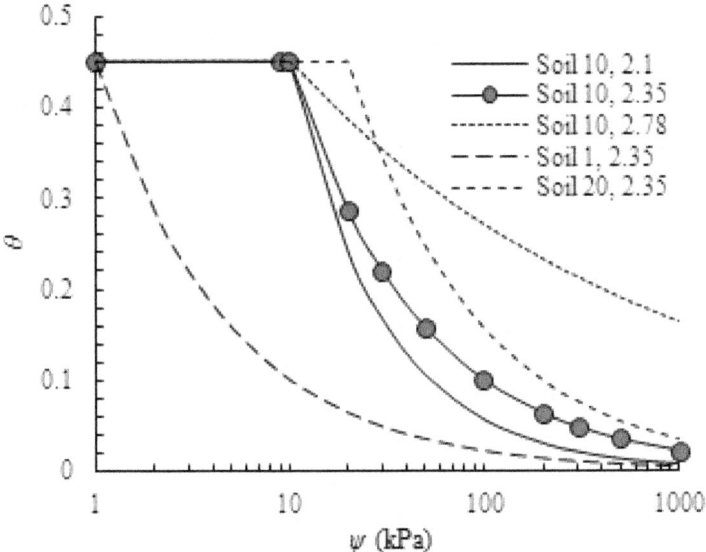

Figure 34: Soil-water characteristic curves for soils

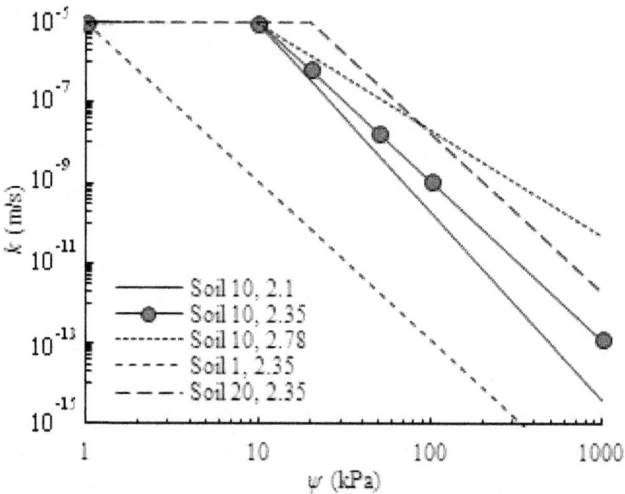

Figure 35: Hydraulic conductivity curves f

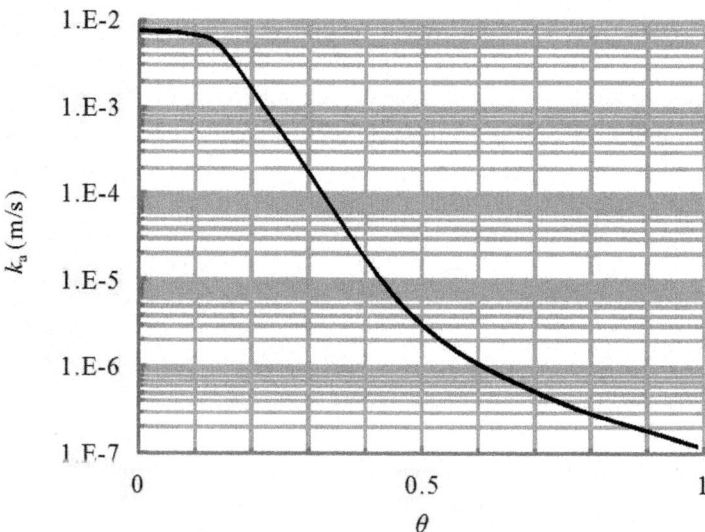

Figure 36: Pore air permeability function used in study

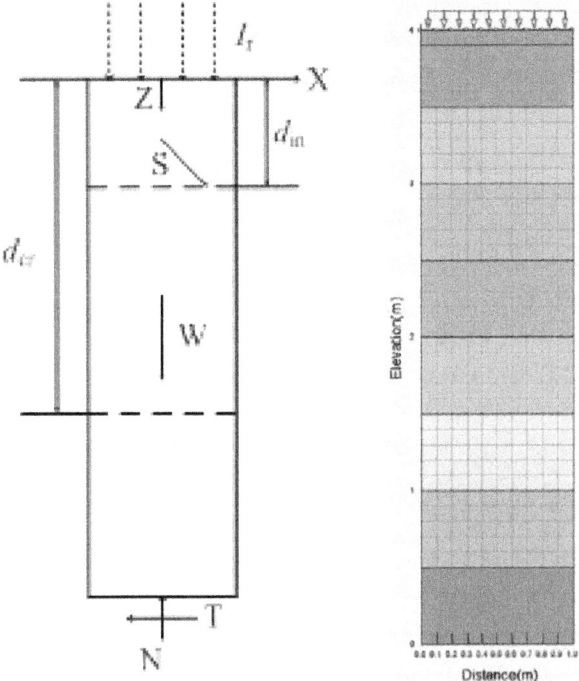

Figure 37: Sketch map of the one-dimensional infiltration model

A 4m-deep soil column was modeled for the one-dimensional analyses, as shown in Fig. 37, and infiltration occurs from the top of the column, while the bottom was set to be a pervious boundary. A linear hydrostatic suction distribution presented for the initial conditions. Analyses for the rainfall infiltration into different types of soils were performed using the unsaturated characteristic curves shown in Figs. 34-36. In applying a top boundary condition to simulate the infiltration of rainfall, it is important to realize that an influx boundary depends on the relationship between the saturated hydraulic conductivity and the rainfall intensity. In the following analyses, if a rainfall intensity greater than or equal to k_s with the non-infiltrating rainfall running off the slope, the top boundary condition is set to a total head equal to 4m; while if field results indicate that the rainfall intensity is less than k_s, then a flux type boundary condition is more appropriate.

Since initial failures often have small depth-to-length ratios and form failure planes parallel to the slope surface, the use of infinite slope analysis in modeling the infiltration process by vertical one-dimensional analysis makes it justified in describing the physical process of failure initiation. However, the methods used in traditional infinite slope analysis must be modified to take into account the variation of the suction profile that results from the infiltration process. There are two distinct failure mechanisms can be initiated by the infiltration process. The failure takes place due to positive pore pressure and it takes place while the suction still exists.

The shear strength of unsaturated soil can be represented by fractal model, and written as follows (Xu, 2004):

$$\tau = c' + (\sigma - u_a)\tan\phi' + \psi_e^{3-D}\psi^{D-2}\tan\phi' \tag{63}$$

where τ is the shear strength; c' is the effective cohesion; $(\sigma-u_a)$ is the total normal stress; ϕ› is effective friction angle; ψ is the suction; ψ_e is the air-entry value. The stability envelope can be represented as (Collins and Znidarcic, 2004):

$$d_{cr} = \frac{c' + \gamma_w h_c \tan\phi' - \gamma_w h_p \tan\phi'}{\gamma\cos^2\beta(\tan\beta - \tan\phi')} \tag{64}$$

where d_{cr} is the critical depth; β is slope angle; h_p is the pore passive pressure and h_c is the pore negative pressure.

We can get suction with analyses of both pore water pressure and pore air pressure in the soil slope under rainfall infiltration. Combining Eq. (63) with Eq. (64), the new stability envelope can be derived and written as follows:

$$d_{cr} = \frac{c' + \psi_e^{3-D}\psi^{D-2}\tan\phi' - \gamma_w h_p \tan\phi'}{\gamma\cos^2\beta(\tan\beta - \tan\phi')} \tag{65}$$

The critical depth for infinite slope failure is now a function of the suction, ψ, pressure head h_p, the given material and slope characteristics, c', ϕ›, γ, ψ_e, ξ, γ_w, and β. In this way, the slope stability issues related to the decrease in shear strength from a loss of soil suction and the development of seepage forces from positive pressure head generation can be clearly understood.

Because the infiltration results and the slope stability results were both presented in terms of the pressure head and the suction profile, the two analyses can be coupled to yield a comprehensive method for determining the location and time of failure for a slope if the soil, slope, and rainfall parameters are given. The methodology of the coupled analysis involves plotting a stability envelope as defined by Eq. (65) for specified soil and slope parameters, over a given infiltration profile generated from the particular rainfall and unsaturated characteristic curves. Coulomb failure and the initiation of slope mobilization are defined at the points where the infiltration trace intersects the stability envelope. Intersections of the infiltration trace and the stability envelope indicate points at which the slope is unstable and can be thought of as "critical depths" of failure. It must therefore be assumed that if the slope is initially stable, the point at which the stability envelope inter sects the initial suction distribution line will not contribute any further information about the failure mechanisms from infiltration.

Here, factors controlling the unsaturated soil slope under infiltration are studied and the results which combine infiltration analyses and stability analyses are shown in Figs.43-47. Infiltration analyses were performed by software SEEP/W coupled with AIR/W in order to calculate the suction file and the stability envelope as defined by Eq.(65) for slope parameters, $\beta=40°$, $\gamma=19.6kN/m^3$, $\gamma_w=10kN/m^3$, and shear strength parameters c'=5kPa, $\phi\rangle=15°$. Other parameters are studied in a systematic way.

Effect of Fractal Dimension

Soil column was saturated due to rainfall infiltration from the initial hydrostatic suction profile. It was shown that infiltration took place in unsaturated soil much faster for the soil with higher fractal dimension. Three fractal dimensions (D=2.1, 2.35 and 2.78) were calculated respectively and the results were shown in Fig.38. It was seen from Fig. 38 that the higher initial hydraulic conductivity existed on the top of soil column which had a higher fractal dimension. On the other hand, the stability envelope for the soil with higher fractal dimension was steeper. There was an obvious inflection point in the stability envelope of the soil with the low fractal dimension. The results also indicated that it was the relative shape of the unsaturated characteristic curves (Fig.34 and Fig. 35) that has a controlling effect on the suction distribution.

Effect of Air-Entry Value

Three air-entry values (ψ_e=1kPa, 10kPa and 20kPa) are calculated respectively and the results as shown in Figs.39. Compared with the results of three different air-entry values, both the infiltration trace and the stability envelope have changed with the different air-entry values. When the air-entry value is relatively small (ψ_e=1kPa), as shown in Fig.39a, passive pore water pressure was generated on the upper soil column at the time t=32h. Due to the low initial permeability on the top of soil column and a state of lower saturation, fewer pores initially filled with water and there are fewer channels available for fluid transport and consequently the flow of water is hindered. With the commencement of infiltration at the top boundary, water is forced into a soil which is not capable of transporting it efficiently, which results in the development of positive pressure heads as the infiltration front progresses downward. While the air-entry value is relatively larger, as shown in Fig.39c, infiltration moves downward much faster and suction has reached to the air-entry value. Passive pore water pressure occurs at the bottom of the soil column which means that there is little unsaturated zone in the middle of soil column during the rainfall infiltration. Besides, the stability envelope is much smooth when the soil has a relatively low air-entry value, and consequently the critical depth is smaller.

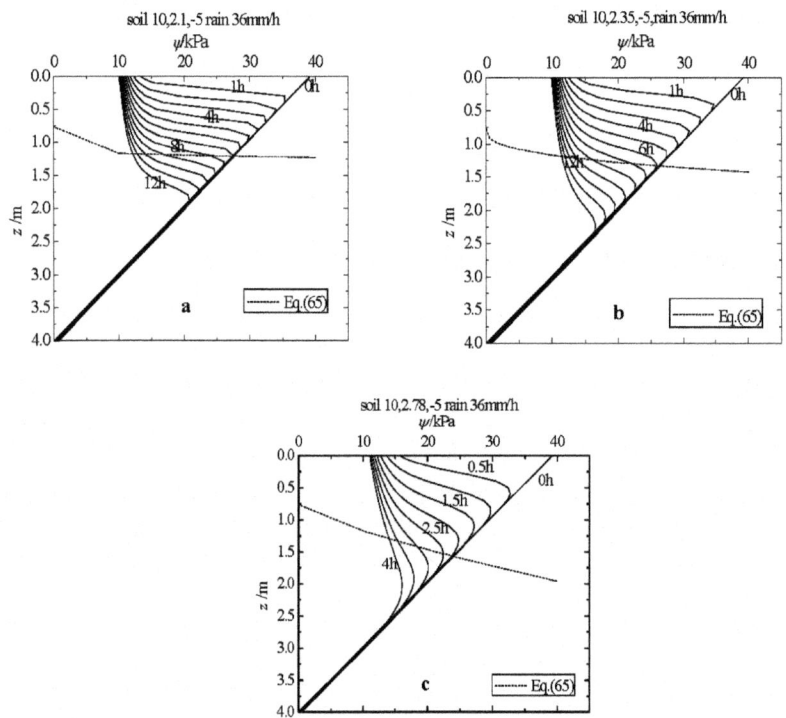

Figure 38: Infiltration results for soils with different fractal dimension and the new stability envelope

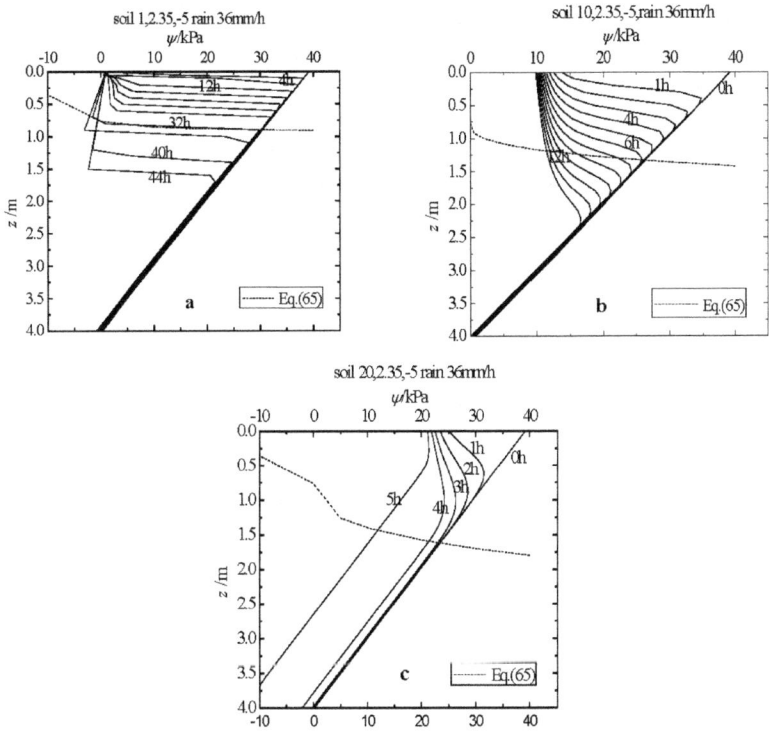

Figure 39: Infiltration results for soils with different air-entry value and the new stability envelope

Effect of Saturated Hydraulic Conductivity

The dominate factors which control the stability of unsaturated soil slope under rainfall infiltration are the soil saturated hydraulic conductivity and rainfall intensity. While the soil saturated hydraulic conductivity determines the water transportation ability of soil and the infiltration quantity most depends on the rainfall intensity. In this parameter study series, three different saturated hydraulic conductivity values ($k_s=1\times10^{-5}$m/s, 1×10^{-6}m/s, 1×10^{-7}m/s) are taken into account and the results are shown in Fig.40.

When the rainfall intensity is much larger than the saturated hydraulic conductivity, soil in the top region of the column has been saturated soon after the infiltration begins and its suction almost reaches to zero at the time t=1.5h. It can be seen that the suction at the base is increasing over time as a positive number, which means that the soil is actually de-saturating at its base while wetting up above. This is caused by the air which becomes trapped when the top has saturated and starts to resist water infiltration. As the situation that the rainfall intensity is much smaller than the saturated hydraulic conductivity as shown in Fig.40a, the suction of the top region decreases with the infiltration developing, however, the top region has remained unsaturated and the infiltration depth is also small due to the small rainfall quantity into the soil.

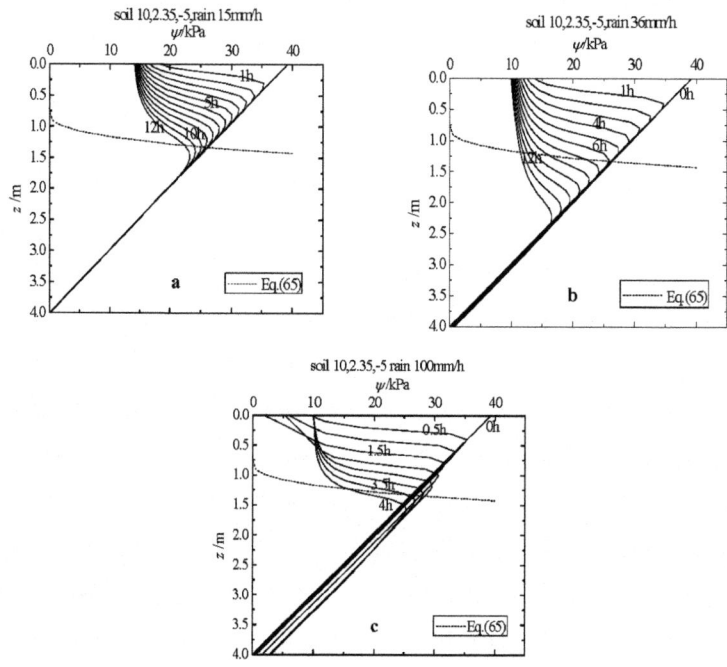

Figure 40: Infiltration results for soils due to different rainfall intensities

Effect of Rainfall Intensity

The rainfall intensity for a certain saturated hydraulic conductivity has taken into considered to simulate the infiltration process. The range is changed from 0.3mm/h to 100mm/h and the relationship between the time of failure and rainfall intensity for the three soil types of saturated hydraulic conductivity has plotted in Fig.41. These curves decrease as exponent forms similarly while the magnitude of different curves has changed dramatically. The time at which the soil slope failure occurs can be about ten hours for the soil saturated hydraulic conductivity k_s equals to 1×10^{-5}m/s, and for the situation that k_s equals to 1×10^{-6}m/s, the time can range from 10 to 100 hours. While much more time is needed for the lower saturated hydraulic conductivity soil type.

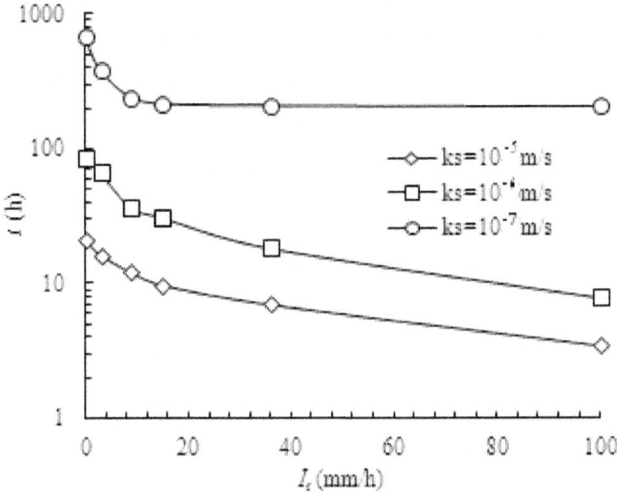

Figure 41: The time of failure vs. rainfall intensity

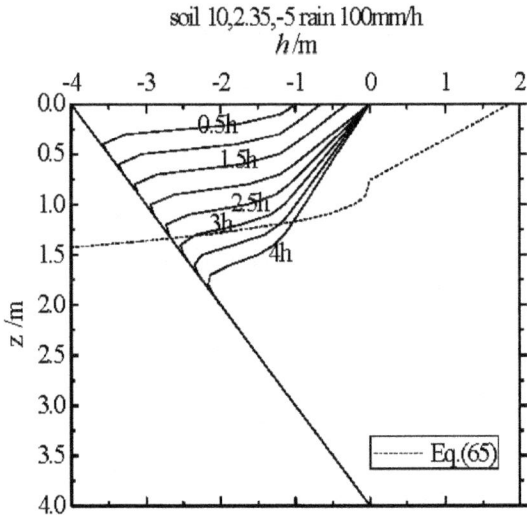

Figure 42: Infiltration results without air phrase consideration

Infiltration Analyses Coupled with the Air Flow Analyses

In the infiltration analyses, air flow movement in the soil pore is also conducted. The effect of the pore air flow movement is especially obvious in the situation when the rainfall intensity is much larger than the soil saturated hydraulic conductivity. Compared with the results of infiltration analyses coupled with air flow analyses, the results without air phrase consideration are shown in Fig.42. In this condition, the negative pore water pressure gradually reaches zero at the top of soil column. While the pore water pressure remains unchanged at the bottom of the column. If we take the pore air flow movement into consideration, the results are different as shown in Fig.40c. Besides, the suction could be calculated more precisely when we attain both the pore water pressure and the pore air pressure by simulating infiltration analyses coupled with air flow analyses.

SLOPE STABILITY ANALYSES DUE TO RAINFALL INFIL-TRATION

It was widely recognized that the rainfall infiltration took a great role in causing landslides, while the relative importance of soil properties, rainfall intensity, initial water table depth and slope geometry in inducing instability of a homogenous unsaturated soil slope under different rainfall was investigated through a series of parametric studies (Rahardjo, et al, 2007). Soil properties and rainfall intensity were found to be the primary factors controlling the instability of slopes due to rainfall, while the initial water table depth and slope geometry only played a secondary role.

The factors affecting the stability of a slope were considered to be the soil properties, rainfall intensity, initial depth of the groundwater table, and the slope geometry (i.e., slope angle and slope height). To assess the effects and relative contribution of controlling factors, a series of parametric studies were performed on a typical geometry of a homogeneous soil slope shown in Fig. 43. It had the boundary conditions as follows: ab, bc, cd=q=I_r(rainfall intensity); ah, de, fg=q=0m³/s (i.e. no flow boundary); and ef, gh=h_t(total head at the side). Four slope heights, Hs,(3m, 4m, 5m and 6m), three slope angles, α,(26.6, 33.7 and 45.0), eight initial depths of groundwater table (GWT), Hw,(0.5m, 1m, 2m, 3m, 5m, 7.5m, 10m and 15m), three fractal dimensions D (2.1, 2.35 and 2.78), four air-entry values ψ_e (1kPa, 10kPa, 20kPa and 50kPa), five values of saturated hydraulic conductivity, k_s (10^{-4}m/s, 10^{-5}m/s, 10^{-6}m/s, 10^{-7}m/s and 10^{-8}m/s) and five rainfall intensities Ir(3mm/h, 9mm/h, 15mm/h, 36mm/h and 100mm/h each for 24h duration) were used in six cases of parametric studies. The shear strength parameters of the soils used in the parametric study were c'=10kPa, effective angle of internal friction,φ›=26°, and unit weight of soil, γ=20kN/m³.

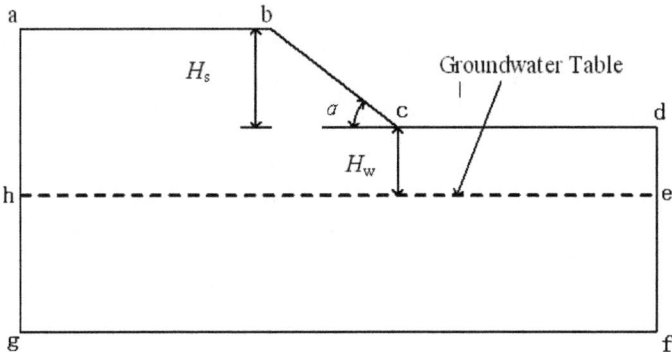

Figure 43: Sketch map of the slope in study

The fractal model of the soil-water characteristic curves for unsaturated soil was written as Eq. (29). The function of relative hydraulic conductivity was used as Eqs. (44) and (45). The derived SWCC and RHC for the all six soils were shown in Figs. 34 and 35. In the legends of these pictures, the symbol S in the soil names represent "soil", the first number means the air-entry value, and the second one means the fractal dimension.

In the seepage analysis, the governing partial differential equation (GEO-SLOPE International Ltd.2007) for a two-dimensional transient water flow used in the finite element seepage model was written as follows:

$$m_w \gamma_w \frac{\partial(H-y)}{\partial t} = \frac{\partial}{\partial y}\left(k_w \frac{\partial H_w}{\partial t}\right) + m_w \frac{\partial P_a}{\partial t} + Q$$

(66)

where H is the total head; kx are ky ars hydraulic conductivity in x, y direction respectively ; Q is the applied boundary flux; m_w is the slope of the SWCC and γ_w is the unit weight of water. Eq. (66) was solved by SEEP/W software (Geo-Slope 2007). The boundary conditions used in the transient seepage analysis were shown in Fig. 43. And the initial condition for the analyses was a hydrostatic condition with a limiting pore water pressure of -75kPa. Then the SEEP/W software generated the negative pore water pressure as the initial condition. The pore-water pressures obtained from the seepage analysis were then used in the slope stability analyses to calculate the factor of the slope safety, F_s.

The equation for unsaturated shear strength was used as Eq. (63). In the study, we chose c'=10kPa,φ'=26°. The Morgenstern-Price Method was adopted in the slope stability analysis. This method considered both shear and normal inter-slice forces, satisfied both moment and force equilibrium, and allowed for a variety of user-selected inter-slice force function. The slope stability analysis using Morgenstern-Price Method was performed using SLOPE/W software (Geo-Slope 2007). The pore-water pressures, u_w, obtained from the transient seepage analyses using SEEP/W were added to SLOPE/W to be incorporated in the slope stability analyses.

The results were presented with attention on the effects of the factors which impacting on the soil slope stability under rainfall for 24h with a combination of various controlling factors.

Effect of Soil Properties

The variation in factor of safety with time for a homogeneous soil slope of constant slope height H_s=6m, initial groundwater table depth H_w=2m, subjected to rainfall intensities of 3, 9 and 15mm/h of respective soil for 24h with a combination of various soil types (different fractal dimensions and air-entry values) were analysized as follows.

Effect of fractal dimension

In Figs.44, the fractal dimension, D was 2, 2.35 and 2.78, respectively. The air-entry value, ψ_e remained 10kPa. It is shown that the reduction of slope stability is larger in the condition of larger fractal dimension under the condition of the same rainfall intensities and the saturated hydraulic conductivity. This means that the soil slope is much safe if the fractal dimension is low while the saturated hydraulic conductivity remains constant. The factor of safety decreases in a short time under rainfall infiltration when the fractal dimension is relatively large, as shown in Fig.44c, while the decrease occurs at a longer time for the low fractal dimension soil type, as shown in Fig.44a.

If the air-entry value and the saturated hydraulic conductivity remain unchanged, the slope of the hydraulic conductivity function is small when the fractal dimension is large. Then the initial hydraulic conductivity value at the beginning of the rainfall infiltration is much big, which results in the speed of rainfall infiltration is great and the saturated degree of soil inside of slope changes quickly leading unsaturated soil to become saturated. The passive pore water pressure generated by the increase of degree of soil saturation makes a negative impact on the instability of soil slope under rainfall.

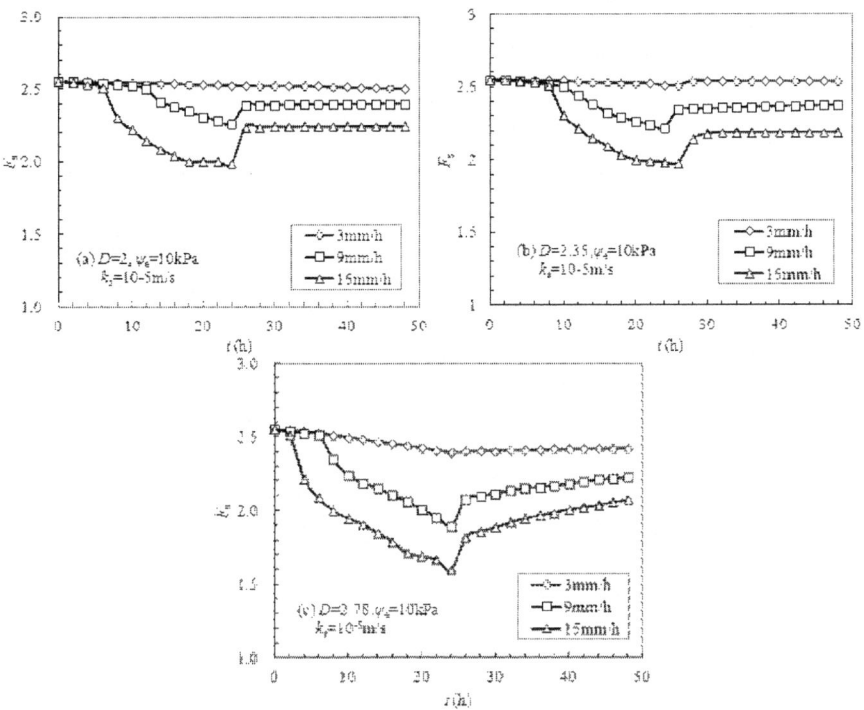

Figure 44: Effect of fractal dimension on slope stability

Effect of air-entry value

The effects of air-entry value on the stability of soil slope due to rainfall infiltration are shown in Figs. 45. The air-entry value, ψ_e is 1kPa, 10kPa, 20kPa and 50kPa respectively and the fractal dimension, Dremains 2.35. The style of the slope safety curve changes a lot with different air-entry values in Figs. 45. In Fig. 45a, the air-entry value, ψ_e is 1kPa and the decrease of factor of safety occurs during 16~20h after rainfall began. As the air-entry value, ψ_e is 10kPa, shown in Fig. 45b, the decrease of slope safety factor occurs at 8h after rainfall began. At the beginning of rainfall infiltration, the decrease happens inFigs. 45c-d, in which the air-entry value is 20kPa and 50kPa respectively. This means that a delay time appears which is depended with the air-entry value. And it is obvious that small air-entry value decides a relative long delay time. Besides, the factor of slope safety remains minimum during another 24 hours after rainfall stops in Fig. 45a, which has a low air-entry value (ψ_e=1kPa). While ψ_e=50kPa, the recovery of safety factor curve in Fig. 45c occurs and dramatically. This is also due to the style of hydraulic conductivity function, as shown in Fig. 35. The initial hydraulic conductivity value becomes large when the air-entry value increases with fractal dimension, saturated hydraulic conductivity and saturation degree remain the same.

Effect of saturated hydraulic conductivity

The effects of saturated hydraulic conductivity on stability of a homogenous soil slope are reflected through the relationship between F_s and k_s, shown in Fig. 46. The saturated hydraulic conductivity, k_s, was varied with five different values of 10^{-8}m/s, 10^{-7}m/s, 10^{-6}m/s, 10^{-5}m/s and 10^{-4}m/s for a homogeneous soil slope of constant soil type S10, 2.35 (ψ_e=10kPa, D=2.35), H_w=2m, H_s=6m, α=33.7°, subjected to rainfall for 24h with six rainfall intensities of 3, 9, 15, 36, 100 and 360mm/h. All the plots in Fig. 47 shows that soil with k_s values of 10^{-8}m/s and 10^{-7}m/s, respectively, are less affected by rainfall. Contrarily, soil with k_s values of 10^{-5}m/s and 10^{-4}m/s are greatly affected by rainfall. It is suggested that soil slopes with a low saturated hydraulic conductivity are relatively safe under short-duration rainfall infiltration and for the soil slopes with a high saturated hydraulic conductivity the stability is affected by the short-duration rainfall greatly. This means that slopes with low saturated hydraulic conductivity need a long-duration rainfall to intrigue the instability.

The F_s(min)-k_s critical curve was presented in the broken line in Fig. 47 as follows:

$$F_s(\min) = \frac{a}{1+be^{-ck_s}}$$

(67)

where F_s(min) is minimum factor of safety; k_s is saturated hydraulic conductivity; a, b and c are the fitting parameters and e is natural number (i.e., 2.71...). Here, a=1.293, b=-0.492, c=1.2×10⁵, and the corresponding coefficient of correlation, r^2 is 0.9998.

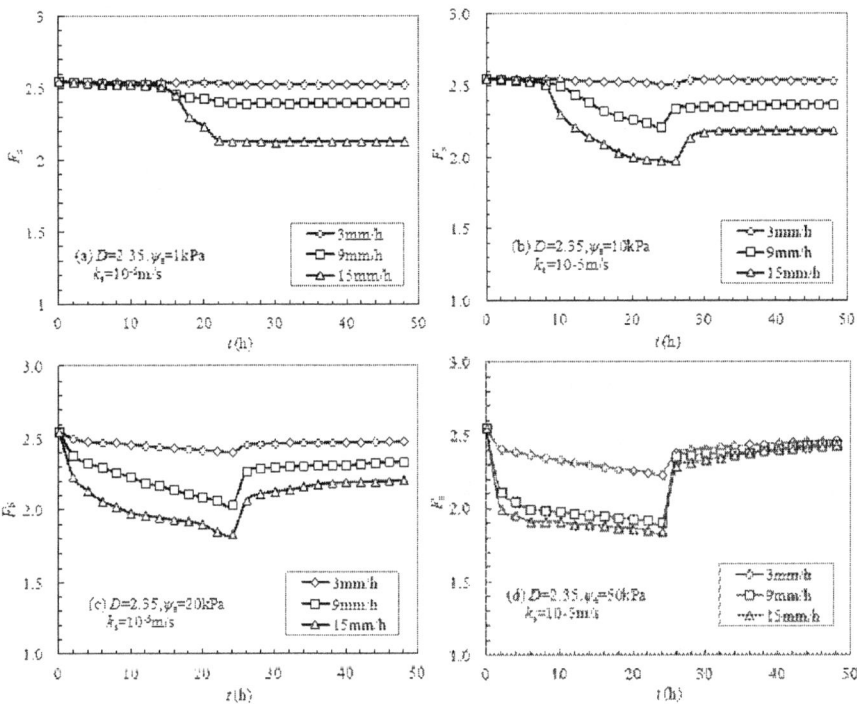

Figure 45: Effect of air-entry value on the slope stability

EFFECT OF RAINFALL INTENSITY

The relationship between minimum factor of safety, F_s(min), versus logarithmic of rainfall intensity, I_r are plotted in Fig. 47. The semi log plot shows that generally the F_s(min) and I_r relationships follow a sigmoid shape in Fig. 47. There are two inflect points for the sigmoid shape line. The Fs(min) is almost constant at very low rainfall intensities for all soil types. And it starts to decrease rapidly when the first inflection point is reached. The trend in F_s(min) versus I_r relationship observed in Fig. 47 can be described by a sigmoid equation (MMF Line) as the form of

$$F_s(\min) = \frac{ab + cI_r^{d}}{b + I_r^{d}}$$

(68)

where F_s(min) is minimum factor of safety; I_r is rainfall intensity; and a, b, c, d are fitting parameters. The values for the fitting parameters and the corresponding coefficient of correlation, r^2, for the sigmoid line in Fig.47.

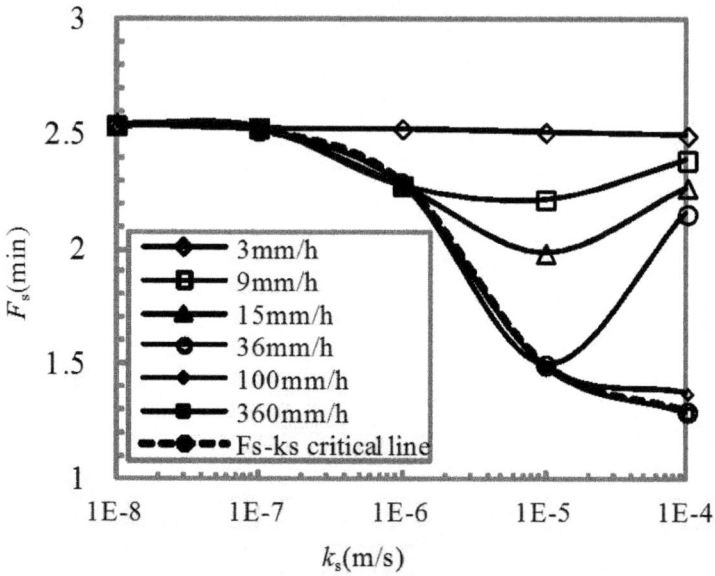

Figure 46: Relationship between K_s and minimum factor of safety

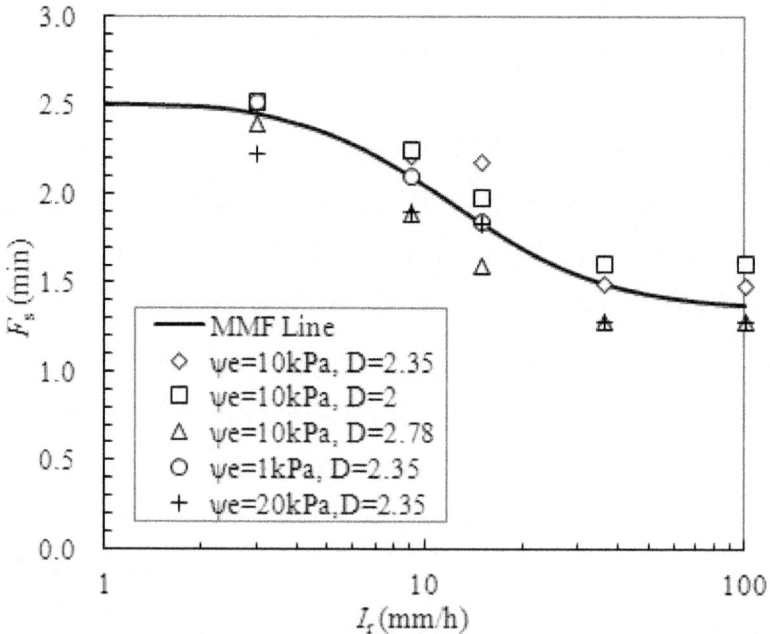

Figure 47: Effect of rainfall intensity on variation of minimum factor of safety

Effect of initial water table location

The effects of initial groundwater table location on stability of a homogenous soil slope are reflected through the relationship between F_s(ini) and F_s(min) with H_w, shown in Fig. 48. The initial depth of water table, H_w, was varied with five different values of 0.5, 1, 2, 2.5, 5, 7.5, 10 and 15m for a homogeneous soil slope of constant soil type S10, 2.35, -5 (ψ_e=10kPa, D=2.35, k_s=10^{-5}m/s), H_s=6m, α=33.7°, subjected to rainfall for 24h with five rainfall intensities of 3, 9, 15, 36, and 100mm/h. The relationship between F_s(ini), and H_w shown in Fig. 48 appears to be linear up to a depth of 7.5m beyond which F_s(ini) remains constant. This is because the initial pore-water pressure profiles generated for the slope at H_w=7.5m as same as those generated when H_w=7.5m. This is due to the limiting pore-water pressure of −75kPa adopted in the analyses. In Fig. 54 it also shows that the rainfall intensity is the dominated factor impacting on the reduction in factor of safety, Fs. And the initial depth of water table, H_w, mainly determines the value of initial factor of safety, F_s(ini). The F_s(ini) is smaller for slopes with a shallower H_w which means that the soil slope stability is much lower. Therefore, slopes with a shallow H_w are more likely to fail due to a rainfall compared with slopes which has a deep H_w.

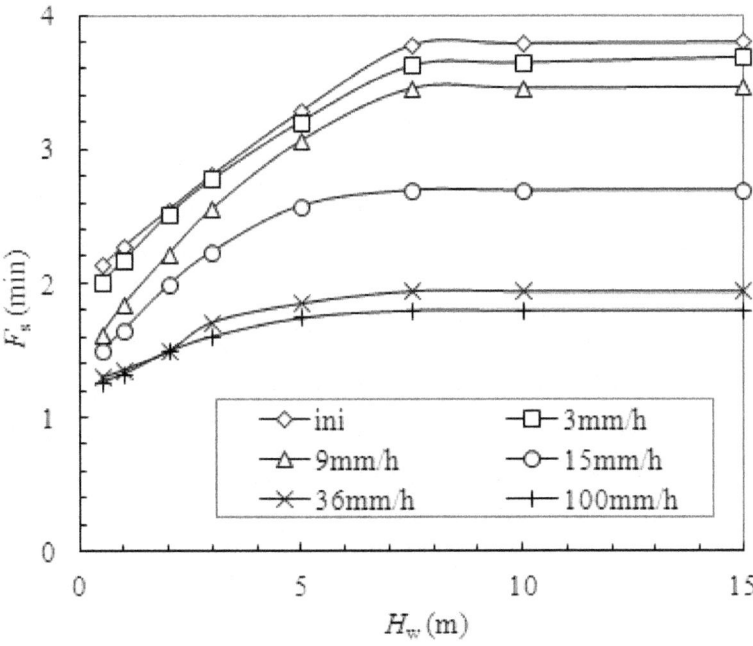

Figure 48: Effect of initial groundwater table depth on factor of safety

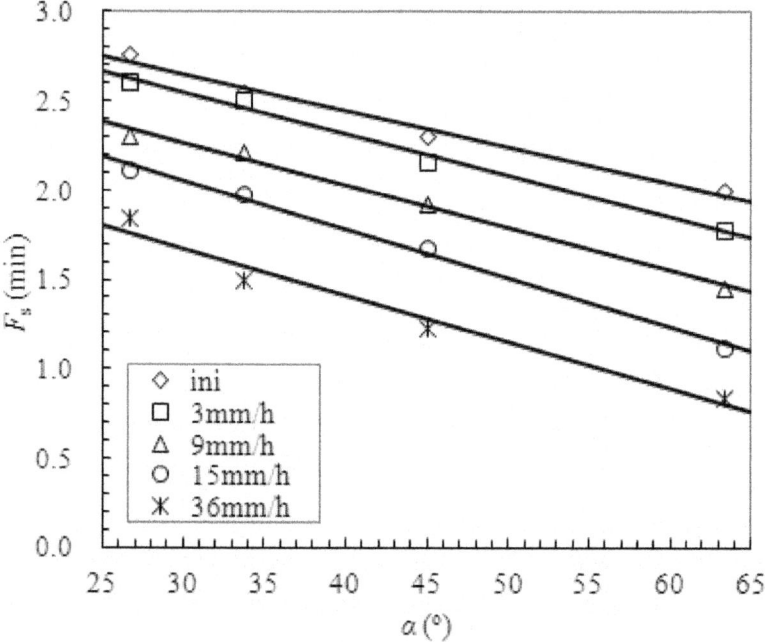

Figure 49: Effect of slope angle on factor of safety

Effect of slope angle

The effect of slope geometry on the stability of a homogenous soil slope is evaluated in terms of slope angle (α) and slope height (H_s). The effects of slope angel on the stability of a homogenous soil slope are reflected through the relationship between $F_s(ini)$ and $F_s(min)$ with α, shown in Fig. 49. The slope angel, α was varied with three different values of 26°, 33.7°, 45°and 63° for a homogeneous soil slope of constant soil type S10, 2.35, -5 (ψ_e=10kPa, D=2.35, k_s=10⁻⁵m/s), H_s=6m, H_w=2m, subjected to rainfall for 24h with five rainfall intensities of 3, 9, 15 and 36mm/h. The relationship between $F_s(min)$ and α shown in Fig. 49 appears to be negative linear. In general, the higher the slope angle, the lower the initial factor of safety and the minimum factor of safety. Because a steep slope will yield a lower factor of safety compared with a flat slope.

Effect of slope height

The effects of slope angel on the stability of a homogenous soil slope are reflected through the relationship between $F_s(ini)$ and $F_s(min)$ with H_s, shown in Fig. 50. The slope height, H_s, was varied with four different values of 3m,

4m, 5m and 6m for a homogeneous soil slope of constant soil type S10, 2.35, -5(ψ_e=10kPa, D=2.35, k_s=10^{-5}m/s), H_w=2m, α=33.7°, subjected to rainfall for 24h with four rainfall intensities of 3mm/h, 9mm/h, 15mm/h and 36mm/h. Fig. 50 shows that initial factor of safety decreases exponentially as the slope height increases. It also suggests that high slopes are generally easier to fail under rainfall due to the low initial factor of safety. The reduction in factor of safety for a high slope is smaller and occurs at a slower rate compared with a low slope.

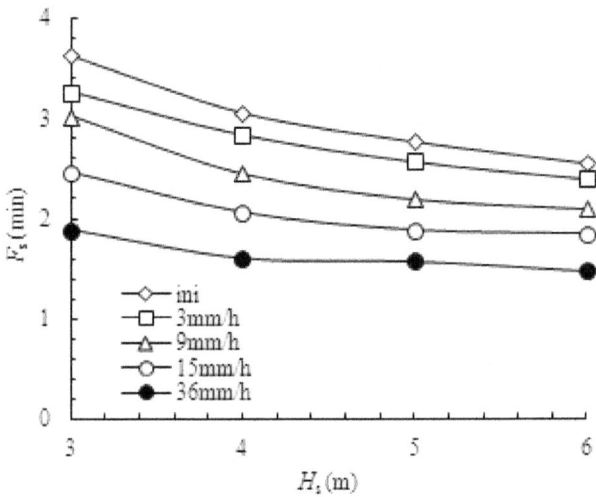

Figure 50: Effect of slope height on factor of safety

ACKNOWLEDGEMENTS

The National Natural Science Foundation of China (Grant No. 41272318) and State Key Laboratory of Ocean Engineering are sincerely acknowledged for their financial support

REFERENCES

1.   Abramowitz M, Stegun I. Handbook of Mathematical Functions. Dover Publications, New York, NY. 1970

2.   Avnir D, Jaroniec M, An isotherm equation for adsorption on fractal surfaces of heterogeneous porous materials. Langmuir, 1989, 5, 1431–1433.

3.   Brooks RH, Corey AT. Properties of porous media affecting fluid flow. ASCE J Irrig Drain Div 1966, 92:61–68.

4.   Brutsaert W. Some methods of calculating unsaturated permeability. Trans Am Soc Agr Engrs 1967;10:400-404.

5.   Brutsaert W. The permeability of a porous medium determined from certain probability laws for pore size distribution. Water Resour Res 1968; 4: 425-34.

6.   Brutsaert W. A concise parameterization of the hydraulic conductivity of unsaturated soils. Advances in Water Resources, 2000, 23, 811-815.

7.   Burdine NT. Relative permeability calculations from pore-size distribution data. Trans Am Inst Min Engrs 1953;198:71-7

8.   Burdine NT. Relative permeability calculation from pore size distribution data. Transactions of the American Institute of Mining Engineers, 1953, 198, 71–78.

9.   Carman PC. Flow of gases through porous media. New York: Academic Press, 1956.

10.   Childs EC, Collis-George N. The permeability of porous materials. Proc Roy Soc A 1950; 201: 392-405.

11.   Collins,B.D., Znidarcic. Stability analyses of rainfall induced landslides [J].Journal of geotechnical and geoenvironmental engineering, 2004,4,362-372.

12.   Corey AT. Pore-size distribution. In: van Genuchten MTh, Leij FJ, Lund LJ, editors. Proceedings of the International Workshop, Indirect Methods for Estimating the Hydraulic Properties of Unsaturated Soils, University of California, Riverside, 1992. p. 37–44.

13.   Crawford JW. The relationship between structure and the hydraulic conductivity of soil. Eur J Soil Sci 1994, 45:493–501.

14.   Delage P, Audiguier M, Cui YJ, Howat MD. Microstructure of a compacted silt. Can Geotech J 1996;33:150-158.

15.   Dirksen C. Unsaturated hydraulic conductivity. In: Smith KA, Mullins CE, editors. Soil analysis physical methods. New York: Dekker; 1991. p. 209–69.

16.   Fatt I, Dykstra H. Relative permeability studies. Transactions of the American Institute of Mining Engineers, 1951, 192: 249-255.

17.   Fuentes C, Vauclin M, Parlange J-I. A note on the soil–water conductivity of a fractal soil. Transp Porous Media 1996, 23:31–36.

18.   Fuentes C, Haverkamp R, Parlange J-Y. Parameter constraints on closed-form soil water relationships. Journal of Hydrology, 1992, 134, 117-142.

19.   Gates JI, Tempelaar-Lietz W. Relative permeabilities of California cores

by the capillary pressure method. Drilling and Production Practice, Am Petrol Inst 1950:285-298.

20.   GEO-SLOPE International Ltd. Seepage Modeling with SEEP/W2007 [M].Third Edition, Canada, 2008,3.

21.   GEO-SLOPE International Ltd. Air Flow Modeling with AIR/W2007 [M].Third Edition, Canada, 2008,3.

22.   GEO-SLOPE International Ltd. Stability Modeling with SLOPE/ W2007 [M].Third Edition, Canada, 2008,3.

23.   Gimenez D, Perfect E, Rawls WJ, Pacheoaky Ya. Fractal models for predicting soil hydraulic properties: a review. Eng Geo 1997;48:161-183.

24.   Hoffmann-Riem H, van Genuchten M.Th, Flühler H. A general model of the hydraulic conductivity of unsaturated soils, in: van Genuchten M.Th, Leij FJ, Wu L. (Eds.), Proceedings of International Workshop, Characterization and Measurements of Hydraulic Properties of Unsaturated Porous Media. Riverside, CA. 22–24th Oct. 1997. University of California, Riverside, 1999, pp. 31–42.

25.   Hunt AG, Gee GW. Application of critical path analysis to fractal porous media:comparison with examples from the Hanford site. Adv in Water Resources 2002;25:129–46.

26.   Irmay S. On the hydraulic conductivity of unsaturated soils. Trans Am Geophys Un 1954;35:463-467.

27.   Jarvis NJ, Messing I. Near-saturated hydraulic conductivity in soils of contrasting texture as measured by tension infiltrometers. Soil Sci Soc Am J 1995;59:27–34.

28.   Kahr G., Kraehenbuehl F, Stoeckli HF, Muller-Vonmoos M. Study of the water –bentonite system by vapour adsorption, immersion colorimetry and X-ray technique: Part II. Heats of immersion, swelling pressures and thermodynamic properties. Clay Miner. 1990, 25, 499–506.

29.   Khaleel R, Relyea JF. Evaluation of van Genuchten–Mualem relationships to estimate unsaturated hydraulic conductivity at low water contents. Water Resour Res 1995; 31: 2659–1668.

30.   Kozeny J. Ueber kapillare leitung des wassers im boden, sitzungsberichte, akad der wissensch. Wien, Math-Naturw Klass Abt IIa 1927;136:271-306.

31.   Mandelbrot BB, The Fractal Geometry of Nature, W.H. Freeman, New York, 1982.

32.   Matsuoka H, Soil Mechanics. Morikita Shuppan Co., Ltd., 1999, p. 61.

33.   Mualem Y. A new model for predicting the hydraulic conductivity of unsaturated porous media. Water Resources Res 1976, 12:513–22.

34.   Polubarinova-Kochina PYa. Theory of ground water movement. Princeton, NJ: Princeton University Press, 1952. p. 613 (translated from the Russian by DeWiest, JMR 1962).

35.   Purcell WR. Capillary pressures-their measurement using mercury and the calculation of permeability therefrom. Trans Am Inst Min Met Engrs Petrol Devel Technol 1949;186:39-46.

36.   Rahardjo H, Ong TH, Rezaur RB. Factors controlling instability of homogeneous soil slopes under rainfall [J]. Journal of geotechnical and geoenvironmental engineering, 2007,12,1532-1543.

37.   Ravichandran N, Krishnapillai S. A Statistical Model for the Relative Hydraulic Conductivity of Water Phase in Unsaturated Soils. International Journal of Geosciences, 2011, 2, 484-492.

38.   Rieu M, Sposito G. Fractal fragmentation, soil porosity and soil water properties. Soil Sci Soc Am J 1991; 55: 1483–1489.

39.   Schaap MG, Leij FJ, van Genuchten MT. Rosetta: A computer program for estimating soil hydraulic parameters with hierarchical pedotransfer functions, J. Hydrol., 2001, 251, 163-176.

40.   Simunek T, Wendroth O, van Genuchten MT, et al., Water Resources Res. 35 (1999) 2965.

41.   Smettem KRJ, Kirkby C. Measuring the hydraulic properties of a stable aggregate soil. J Hydro 1990, 117: 1-13.

42.   Stingaciu LR, Weihermüller L, Haber-Pohlmeier S et al. Determination of pore size distribution and hydraulic properties using nuclear magnetic resonance relaxometry: A comparative study of laboratory methods. Water Resources Research, 2009, 46: W11510.

43.   Sugii T, Uno T, Hayashi T, Proceedings of the 31st Japan National Conference on Geotechnical Engineering, Kitami, Japan, 1996, pp. 2075

44.   Toledo PG, Novy RA, Davis HT, Scriven LE. Hydraulic conductivity of porous media at low water content. Soil Sci Soc Am J 1990;54:673–9.

45.   Touma J, Vauclin M. Experimental and numerical analysis of two-phase infiltration in a partially saturated soil. Transport in Porous Media, 1986, 1, 27-55.

46.   Touma J. Comparison of the soil hydraulic conductivity predicted from its water retention expressed by the equation of van Genuchten and different capillary models. European Journal of Soil Science, 2009, 60:

671-680

47.   Tyler SW, Wheatcraft SW. Fractal process in soil water retention. Water Res Res 1990, 26:1047–54.

48.   Uno T, Kamiya K, Tanaka K. The distribution of sand void diameter by air intrusion method and moisture characteristic curve method. Proc Japan Soc Civil Engrg, 1998, 603(III-44): 35-44.

49.   van Genuchten MTh, Leij FJ. On estimating the hydraulic properties of unsaturated soils. In: van Genuchten M.Th, Leij FJ, Lund LJ, editors. Proceedings of the International Workshop, Indirect Methods for Estimating the Hydraulic Properties of Unsaturated Soils, University of California, Riverside, 1992. p.1-14.

50.   van Damme H. Scale invariance and hydric behaviour of soils and clays. CR Acad Sci Paris 1995, 320: 665–81.

51.   van Genuchten MTh. A close form equation for predicting the hydraulic conductivity of unsaturated soils. Soil Sci Soc Am J 1980;44:892–7.

52.   Vogel HJ, Roth K. A new approach for determination effective soil hydraulic function. Eur J Soil Sci 1998;49:547–56.

53.   Watabe Y, Leroueil S, Le Bihan J-P, Influence of compacted conditions on pore-size distribution and saturated hydraulic conductivity of a glacial till, Can Geotech J, 2000, 37:1184-1192.

54.   Wheeler SJ, Karbue D, Constitutive modeling, In: Alonso EE, Delage P, Proc. 1st Int Conf Unsat Soils Rotterdam: AA Balkema, 1996.

55.   Wyllie MRJ, Spangler MB. Application of electrical resistivity measurements to problem of fluid flow in porous media. Bull Am Ass Petrol Geologists 1952, 36:359-403.

56.   Xu YF, Sun DA. A fractal model for soil pores and its application to determination of water permeability. Physica A 2002;316(1-4): 56–64.

57.   Xu YF, Dong P. Fractal approach to hydraulic properties in unsaturated porous media. Chaos, Solitons & Fractals, 2004, 19(2): 327-337

58.   Xu YF, Sun DA, Yao YP. Surface fractal dimension of bentonite and its application to determination of swelling properties. Chaos, Solitons & Fractals, 2004, 19(2): 347-356.

59.   Xu YF, Calculation of unsaturated hydraulic conductivity using a fractal model for the pore-size distribution, Computer and Geotechnics, 2004, 31(7):549-557.

60.   Xu YF. Fractal approach to unsaturated shear strength. Journal of Geotechnical and Geoenvironmental Engineering, 2004, 130(3): 264-273

Chapter 9

EQUIVALENT POROUS MEDIA (EPM) SIMULATION OF GROUNDWATER HYDRAULICS AND CONTAMINANT TRANSPORT IN KARST AQUIFERS

Reza Ghasemizadeh[1], Xue Yu1, Christoph Butscher[2], Ferdi Hellweger[1], Ingrid Padilla[3], Akram Alshawabkeh[1]

[1] Department of Civil and Environmental Engineering, Northeastern University, Boston, Massachusetts 02115, United States of America,

[2] Department of Engineering Geology, Institute of Applied Geosciences, Karlsruhe Institute of Technology, 76131, Karlsruhe, Germany,

[3] Department of Civil Engineering and Surveying, University of Puerto Rico, Mayaguez, Puerto Rico 00682, United States of America

ABSTRACT

Karst aquifers have a high degree of heterogeneity and anisotropy in their geologic and hydrogeologic properties which makes predicting their behavior difficult. This paper evaluates the application of the Equivalent Porous Media (EPM) approach to simulate groundwater hydraulics and contaminant transport in karst aquifers using an example from the North Coast limestone aquifer system in Puerto Rico. The goal is to evaluate if the EPM approach, which approximates the karst features with a conceptualized, equivalent continuous medium, is feasible for an actual project, based on available data and the study scale and purpose. Existing National Oceanic Atmospheric Administration (NOAA) data and previous hydrogeological U. S. Geological Survey (USGS) studies were used to define the model input parameters. Hydraulic conductivity and specific yield were estimated using measured groundwater heads over the study area and further calibrated against continuous water level data of three USGS observation wells. The water-table fluctuation results indicate that the model can practically reflect the steady-state groundwater hydraulics (normalized RMSE of 12.4%) and long-term variability (normalized RMSE of 3.0%) at regional and intermediate scales and can be applied to predict future water table behavior under different hydrogeological conditions. The application of the EPM approach to simulate transport is limited because it does not directly consider possible irregular conduit flow pathways. However, the results from the present study suggest that the EPM approach is capable to reproduce the spreading of a TCE plume at intermediate scales with sufficient accuracy (normalized RMSE of 8.45%) for groundwater resources management and the planning of contamination mitigation strategies.

INTRODUCTION

Karst aquifers account for 25% of groundwater resources in the world and 40% in the US [1]. They are formed when the dissolution process in primarily soluble carbonate rocks creates complex networks of preferential flow pathways, such as solutionary fractures and conduits, within the rock matrix (karstification). Conduits are crucial for groundwater flow and contaminant transport in karst aquifers [2], but their distribution is often unknown, thus limiting the applicability and validity of the numerical models that require detailed data on conduits [3]. Subsurface flow within the aquifer ranges from laminar to turbulent, with laminar flow in the rock matrix and predominantly turbulent flow in conduits, depending on flow velocities [4]. Karst areas include swallets, sinkholes, infiltrating streams, and other highly porous surface features that limit the availability of surface water, making groundwater the primary water resource for domestic, agricultural, and industrial utilization.

Traditional simulation of groundwater hydrodynamics with numerical models based on Darcy's law may not be directly applicable for modeling flow in karst [5, 6]. Such models are typically used for laminar groundwater flow regime and slow groundwater velocity conditions and their application in karst aquifers require extra attention. In an Equivalent Porous Media (EPM) approach for karst groundwater systems, the default assumption is that carbonate aquifers' behavior is equivalent to porous media for both flow and transport. Also known as single continuum porous equivalent approach (SCPE), heterogeneous continuum approach, smeared conduit approach, or single continuum approach [7, 8, 9, 10], it is the simplest distributed modeling approach for karst aquifers. For the reasons outlined above, its ability to simulate groundwater flow in karst, however, is limited.

The EPM approach assumes that the rock matrix including fractures and conduit networks can be represented by an equivalent porous medium with equivalent hydraulic conductivity in a certain area [11, 12]. In highly karstified aquifers, however, the contaminant transport may depend primarily on the karst conduit network rather than matrix hydraulic conductivity. EPM models often do not distinctly account for preferential flow; instead, they approximate the overall local conductivity of the matrix as well as possible fractures and conduit networks with an enhanced equivalent conductivity [5, 13]. Despite this limitation, it can lead to representative results depending on the degree of aquifer karstification and the scale of the modeling effort. Generally, the EPM approach is more suitable for regional scales rather than local and intermediate scales [5, 14].

Scanlon et al. [5] evaluated the accuracy of two different EPM approaches, lumped parameter and distributed parameter, for simulating regional scale (330 km²) groundwater flow in the highly karstified Barton Springs Edwards aquifer in Austin, Texas. Both models simulated the temporal variations in spring discharge fairly accurate, but the latter approach reproduced the effect of pumping on groundwater levels, evaluated the potentiometric surface at different times, and spring discharge more accurately. According to their study, it is practical to simulate karstic aquifers as equivalent porous media for groundwater flow but not for contaminant transport (especially at local scales) or to delineate protection zones for wells or springs because transport velocities may be substantially underestimated.

Using both porous medium and conduit network simulations, Worthington [6] investigated high-permeability interconnected conduits in the limestone aquifer at Mammoth Cave (Kentucky, USA) at a regional scale (area of 258 km²) both experimentally and numerically to explore if the aquifer behaves more similar to a porous medium or to a karst aquifer with flow both in the rock matrix and karst conduits. While the underground drainage pattern was comparatively well-known from previous tracer studies, that study identified a significant difference in the simulated aquifer behavior with and without considering conduits; especially at local scales, the EPM approach was inadequate to simulate solute transport, represent spring hydrographs, and to interpret tracer test results.

Using MODFLOW 2000, Putnam and Long [15] developed a regional EPM flow model for the Madison and Minnelusa aquifers (area of 2590 km²) to synthesize available hydrogeological data and quantify the regional water budget. Saller et al. [16] further developed the above model by including high conductivity zones, representing karst conduits in the EPM model, which resulted in a better match with the observed transient well data; however, the simulated spring discharge was not improved and was similar to the original EPM.

If sufficient geologic data are available so that the location and characteristics of conduit networks within karst aquifers are fairly well known, some investigators have used MODFLOW to simulate those networks (on a scale order of less than 100 m to several kilometers) with various techniques, including lines of drain cells and internal sinks [17, 18] and lines of high hydraulic conductivity cells [19]. Although not directly comparable, results from above studies may be applicable to other karst aquifers on a case-by-case basis. Yet there remain questions as to how reliable and to what temporal resolution the EPM approach can replicate groundwater level fluctuation in karst or whether EPMs can simulate the long-term fate and transport of point source contamination.

The present study evaluates the utility of the Equivalent Porous Media (EPM) approach to simulate dynamic groundwater hydraulics and contaminant transport in highly permeable karst unconfined aquifers using an example from the North Coast limestone aquifer system in Puerto Rico at an intermediate scale (area of 132.5 km^2 and scale order of 10 kilometers). The three dimensional finite difference model MODFLOW 2005 [20] was used for simulating flow, and the transient transport model MT3DMS [21] was used to simulate the spreading of a point source TCE contamination at a Superfund site. The main goal of this study is to evaluate if the EPM approach is suitable and practical for the assessment of major contaminated sites in karst areas, based on available data and the study's scale and purpose.

MATERIALS AND METHODS

Ethics Statement

No specific permits were required for the described field studies, nor for the locations and activities. The studied location is not privately-owned or protected in any way, and the field studies did not involve endangered or protected species.

Study Site

The study was carried out at a test site within the north coast region of Puerto Rico. Karst limestone features cover 27.5% of the land surface of Puerto Rico (Fig 1). The Northern Karst Belt covers an area of approximately 2000 km^2 (16% of the island) and elevates at heights ranging from about 300 m in the mountainous areas to sea level at the coast (Fig 1) [22]. This aquifer system is the largest aquifer of the island and contains the island's most extensive freshwater aquifer, accounting for 80% of northern Puerto Rico's groundwater resources [23].

Figure 1: Location of study site in northern Puerto Rico. The map includes study area boundary, municipalities, major streams, focus area, Industrial Park, land surface elevation, and position of cross-section A-B. doi:10.1371/journal.pone.0138954.g001

Geology and hydrology.

The North Coast limestone aquifer system of Puerto Rico is sub-divided into four generalized hydrogeological units: the upper water-table aquifer, the middle confining unit, the lower confined aquifer, and the basal confining unit [24]. The thickness of the karst aquifer system increases from the south (Southern Karst Uplands) to the central part of the Northern Karst Belt (Karst Upland Plateau), and subsequently decreases to the coastal plane in the north (c.f.Fig 2).

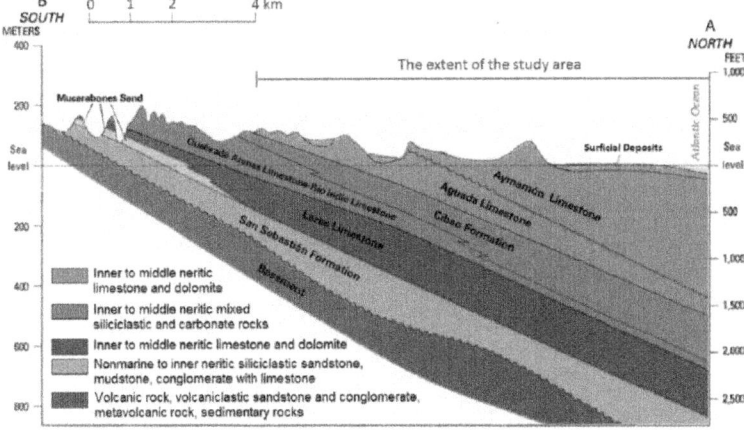

Figure 2: Geologic formations of the North Coast Aquifer system. The geo-

logic sections are along cross-section A-B (see Fig 1). doi:10.1371/journal. pone.0138954.g002

The aquifer system consists of a series of Tertiary limestone formations containing eogenetic karst [25, 26, 27]. The term eogenetic refers to the early stage of karst porosity evolution at low burial, involving surface processes and meteoric diagenesis [28, 29]. The limestone formations include, from north to south and stratigraphically from top to bottom, the Aymamon limestone, the Aguada limestone, the Cibao Formation, and the Lares limestone (Fig 2).

The Tertiary limestone units are locally covered with Quaternary soils such as clay, blanket sand deposits, marine terrain deposits and stream valley alluvium [30, 31]. In the area of Vega Alta, where the present study was carried out, the aquifer system does not contain the Lares limestone formation. Due to the high annual rainfall and the warm climate of the area (average annual temperature about 25°C), the limestone units and soils are highly prone to erosional processes, and climatic conditions are favorable for karst development. The karst conduit systems typically occupy only a small portion of the total aquifer porosity [9]. Nevertheless, the karst conduit system may have a major impact on the hydraulic behavior of the total karst system. The frequency of sinkholes is < 1% in most locations and is much less in Vega Alta than in other municipalities of the north coast. As a result, point recharge into the aquifer is minor, leading to relatively low groundwater level fluctuations [32].

The limestone formations in northern Puerto Rico are slightly or moderately karstified [33]. Generally, the hydraulic conductivities of these units decrease exponentially from a maximum of about 2,040 m/day in the upper Aymamon limestone to a minimum of about 0.04 m/day in the basal Lares limestone [32]. In Vega Alta area, a first set of horizontal hydraulic conductivities (K values) and their vertical distribution over the model layers were defined by Sepúlveda [34] based on slug tests data conducted in 23 multiport wells with different monitoring ports at various depths. The slug tests indicated that K values ranged from 0.06 to 305 m/d, with generally higher values near the water table. The karst upland plateau in Vega Alta encompasses numerous closely spaced deep sinkholes and few gentle depressions in Aguada limestone and Aymamon limestone, respectively. Hydrological investigations [32, 35] demonstrate highly variable porosity and high permeability contrasts in short distances reflecting the variable distribution of conduit porosity in the north coast and suggest that water-bearing conduits are present between Rio Camuy and the Aguadilla area. The presence of caverns in test wells drilled in Hatillo and Isabela in the north and northwest of Puerto Rico has been verified by Rodriguez and Hartley [36]. Using aerial photographs and geophysical methods, Rodriguez and Richards [37] investigated a few sites in northwestern Puerto Rico to detect the presence of conduits and probability of conduit-controlled groundwater flow based on presence of natural potential anomalies. The exact locations of highly transmissive features such as cavities, conduits and fractures in central northern aquifers, however, are uncertain; and no dye tracer or cave diving evidence exists for locations and properties of subsurface conduits discharging groundwater into the northern coastal plain or the submarine outcrop of the aquifer [38]. Rodriguez [39] classified the springs in the north coast limestone based on their hydrodynamics and suggested that some springs are interconnected by an integrated conduit network. Mackovic spring (with limited discharge data) in Vega Alta exhibits little short-term response to rainfall, while Maguayo spring (with flow range of 147 to 2,789 m3/d) in Dorado exhibits a strong short-term response to rainfall and therefore they are known as diffuse-flow and conduit-type minor springs, respectively (see Fig 3).

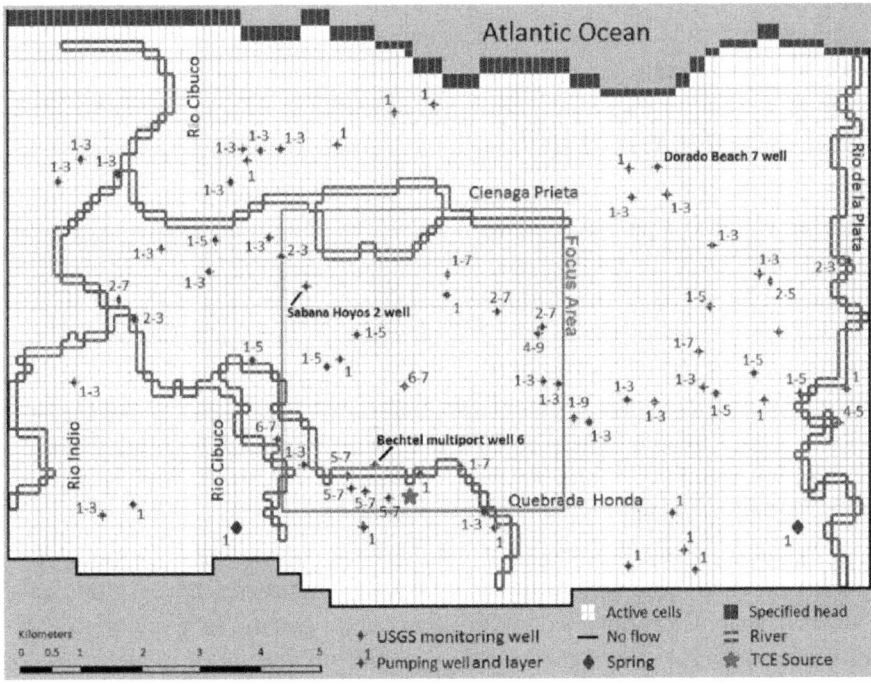

Figure 3: The model domain and defined boundaries. The domain extends beyond the two rivers and consists of focus area, location of TCE source, springs, USGS monitoring wells, and groundwater extraction wells. The well labels indicate the range of layers from which the groundwater was extracted. doi:10.1371/journal.pone.0138954. g003

The rainfall in Puerto Rico varies geographically and seasonally, with February being the driest month and September and October the wettest months [40]. There is a dry period starting in December and usually ending in March or April, followed with a spring rainfall period in April and May, an erratic, semidry period in June and July, and a wet season from August through November [33]. The average annual rainfall in the area of the North Coast limestone aquifer system is about 180 cm (NOAA). With a tropical climate and a high average level of solar radiation, the area experiences high evapotranspiration and receives relatively low rainfall intensity and duration, especially in winters. The average annual evapotranspiration is about 110 cm, or 63% of the average annual rainfall [33]. In Vega Alta area, assuming a minimal runoff, a preliminary estimate of annual aquifer recharge rates for watersheds in Puerto Rico was obtained by Giusti [33] from a correlation of evapotranspiration vs. rainfall (up to 84 cm recharge for a mean annual precipitation of 200 cm). Gómez and Torres-Sierra [41] estimated the spatial distribution of recharge for Vega Alta aquifer and developed a 2-dimensional groundwater model to assess the impact of additional withdrawals on groundwater balance.

Contamination concerns.

The focus area of the present study was the unconfined Vega Alta limestone aquifer, which is a part of regional North Coast limestone aquifer. The aquifer consists of Aymamon and Aguada limestone (the two top layers in Fig 2) and is underlain by Cibao Formation in the south and saltwater-freshwater interface in the center and to the north. It is located in the Vega Alta and Dorado municipalities in the central part of the north coast of Puerto Rico between the municipalities Vega Baja and Toa Baja. The study area is bounded by the USGS Vega Alta quadrangle and the low permeability Cibao Formation to the south, the Atlantic Ocean coast to the north, the Rio de la Plata River to the west, and the Rio Indio River to the east (Fig 1).

Because of the high aquifer productivity, among other reasons, many pharmaceutical, chemical, and manufacturing industries have settled in the North Coast of Puerto Rico, with subsequent growth in population. Many of these industries rely on the use of hazardous materials, which can enter the karst groundwater from accidental spills and deliberate disposals. Many sinkhole depressions have also been used as clandestine waste disposal sites. Water quality surveys in Puerto Rico have shown a vast contamination of the North Coast limestone aquifer [42]. Contamination with chlorinated chemicals, which have been measured in a large percentage of sample wells, is of great concern. Extensive contamination has resulted in the closure of 41%

of drinking water supply wells by 1987 [43]. Since then, there have been more closures, and recently, groundwater has been avoided as a primary source of public water supply in northern Puerto Rico [23].

Up to 49 thousand cubic meters per day of groundwater was extracted in 1991 from 20 public supply wells in Vega Alta and Dorado for public water supply and private usage (industrial, commercial, and agricultural) [40]. Being among the most productive aquifers of the island, this aquifer is at the same time one of the most contaminated. Volatile Organic Compounds (VOCs) used in the metal, electronic, and dry cleaning industries have been leaching from beneath an Industrial Park (see Fig 1) for decades. It was first considered contaminated in 1983 when water samples from 17 of 90 wells collected at the Vega Alta Public Supply Wells site (Latitude: 18.41806, Longitude: -66.33028) were identified to contain high concentrations of methylene chloride, extractable organic compounds, trichloroethylene (TCE) and other VOCs [44].

The US Environmental Protection Agency (EPA) was concerned about health effects of this contamination, and the contaminated wells were immediately shut down by the Puerto Rico Aqueduct and Sewer Authority (PRASA) being responsible for the operation and maintenance of the public water supply system in Vega Alta. In 1984, the site was included in the Superfund National Priorities List (NPL) as "Vega Alta Public Supply Wells" Superfund site. The Vega Alta landfill and the Industrial Park have been recognized as possible sources for TCE and other VOCs. The Vega Alta limestone aquifer was estimated to contain 5900 kg (in 1990) and 5800 kg (in 1992) of TCE, and solute TCE influx into the aquifer was estimated at 9.98 kg/yr under long term average net recharge rates [22]. The drinking water supply of about 40,000 people in Vega Alta municipality depends upon the health of the highly vulnerable karst aquifer. Establishing a thorough understanding of the hydrogeology of this major karst aquifer system is important for water resources and groundwater management and also provides the foundation for predicting the contamination migration.

Model Setup and Calibration

To illustrate the application, performance, and limitations of the EPM modeling approach in evaluating the release and transport of contamination in karst aquifers, a MODFLOW model was developed and paired with MT3DMS; both were implemented in Visual MODFLOW Standard 2011.1 [45]. The variably-spaced grid is composed of 76 rows and 113 columns, encompassing a total area of 132.5 km^2 (Fig 3). The grid is oriented south-north in order to align the model columns with the dominant coastward direction of groundwater flow. The rectangular cells have a maximum size ($\Delta X \times \Delta Y$) of 236×248 m^2 and a minimum size of 118×124 m². The final grid consists of 8588 cells in each layer, with the finest refinement around the contamination source (the eastern side of the Industrial Park) and water extraction wells. The unconfined aquifer was subdivided into 12 layers with basis elevations of -3.05, -15.24, -22.86, -30.48, -38.1, -45.72, -53.34, -60.96, -68.58, -76.2, -86.87, and -97.54 m a.s.l, whereas the last elevation corresponds to the base of the karst aquifer (lower model boundary). The subdivision is based on similar hydraulic conductivities and initial TCE concentrations in these layers [46]. The saturation thickness of the top layer varies as a function of the water table.

Transient simulations are needed to analyze time dependent variables such as the response of water table in aquifer systems to transient signals like seasonal precipitation pulses. The transient flow simulations are run using monthly rainfall between February 1983 and December 2011 (29 years), and also using a refined daily rainfall between June 2004 and December 2005 (19 months) under the same conditions to evaluate the performance of the flow model in a greater temporal resolution.

The Atlantic Ocean coastline in the north is defined as a specified (constant) head boundary of the model with zero head at the sea-level. Predominantly south to north groundwater flow, suggested by historical water level data, implies no-flow boundary conditions along the western and eastern boundaries (c.f., Fig 3). These assumptions are consistent with previous modeling efforts [34, 41, 46] and are constant throughout the simulations. The unconfined aquifer bottom is underlain by the Cibao formation toward the south and by a static saltwater interface (groundwater with estimated dissolved solids exceeding 35,000 mg/L) toward the northern part of the area. In the Vega Alta karst system, the southern Cibao formation is almost impermeable with very low effective porosity due to the existence of clay layers and the absence of karst features. Therefore, a no-flow boundary for each layer is used as the southern and bottom boundary of the model, as well as the northern model boundary (Atlantic Ocean) in all layers except for the topmost two layers in the north.

Recharge is a key component of the hydrological balance in groundwater systems, and it is often difficult to quantify this parameter since it depends on variety of complex factors such as runoff and evapotranspiration. Due to the relatively low relief (maximum altitude of around 150 m a.s.l.) in the Vega Alta area, the average rainfall depends only little on the altitude (slightly greater values at higher elevations), which is confirmed by NOAA data. Therefore the model assumes that the rainfall is independent of the elevation, and rainfall data are only taken from Dorado rain gauge in the study area for model setup. Diffuse infiltration of rainfall through permeable outcrops is the major source of recharge to the aquifer. The net transient recharge is spatially distributed in six zones over the aquifer with rates calculated by multiplying the monthly rainfall to the calibrated discrete aerially-weighted conversion factors of 3%, 6%, 9%, 16%, 25%, and 31% of the net rainfall for the different geologic formations. The highest recharge rates correspond to the limestone outcrops and medium and lower recharge rates for areas with thick blanket deposits and coastal clay deposits, respectively. Aerial recharge was considered as specified flux (Neumann) boundary condition.

The major streams flowing through the study area are Rio de la Plata (the longest in the island) and Rio Cibuco with drainage areas of 227 and 519 square kilometers, respectively (c.f., Fig 1). At base flow conditions, river stages measured at the intersections with highway 2 correspond to 1.52 and 0.91 m a.s.l respectively. Both water level stages are assumed constant. The rivers were implemented as a transfer (Cauchy) boundary condition in the topmost layer of the model, with riverbed conductance values used for river cells ranging from 4.645 to 371.61 m^2/day, according to the studies of Monroe [30] and Gómez and Torres-Sierra [41].

There were 72 private and public pumping wells in the Vega Alta study area. Historically, well closure was the major reason for the decline in groundwater withdrawals, and the lost water public supply was compensated from surface-water sources [47]. Withdrawal rates and filter depth data were taken from field measurement by Sepúlveda [34] and PRASA before 1992, and were assumed constant between 1992 and 2011. The transient model requires initial conditions for the hydraulic head and contaminant distribution. A map of water levels measured in February 1983 [46] is used to define initial conditions of the transient flow model. The effects of uncertainty in initial conditions on simulation results decrease as the simulation progresses. Therefore, simulations were given sufficient "spin-up" time before the period being modeled to limit errors associated with possibly erroneous initial conditions.

Measured water table heads of February 1983 [48] in 70 wells were used for steady-state calibration to estimate the initial conditions for subsequent transient simulation. Parameters used for steady-state flow calibration are net recharge and hydraulic conductivity. These values are slightly adjusted and refined during the manual steady-state calibration processes to match modeled and spatially observed hydraulic heads (Fig 4).

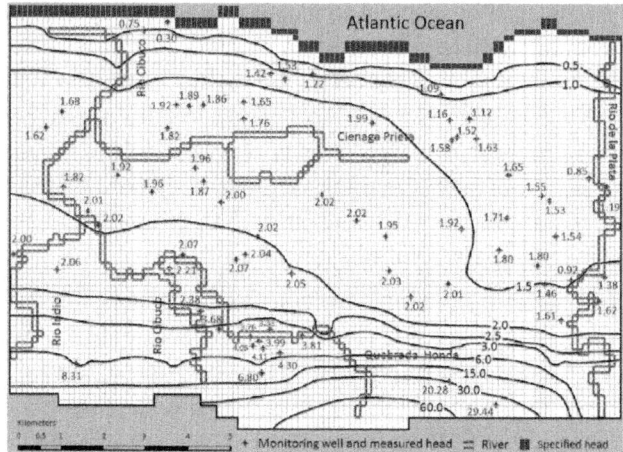

Figure 4: Calibrated steady-state water table contours for February 1983. The simulated groundwater levels were used as the initial conditions for the transient flow model. The monitoring well labels represent the measured groundwater heads (m above sea level). doi:10.1371/journal.pone.0138954.g004

Transmissivity within the upper limestone aquifer is largely regulated by the thickness of the freshwater lens that thins landward and coastward [24]. The distribution of hydraulic conductivities first estimated by Sepúlveda [34] were further adjusted for each layer in this model (Fig 5A), with lower values in the Cibao formation and higher values in the Aguada and Aymamon formations. The highest K values in the middle of model domain correspond to the upland plateau characterized by outcrops and closely spaced sinkholes which are surrounded by low rolling hills.

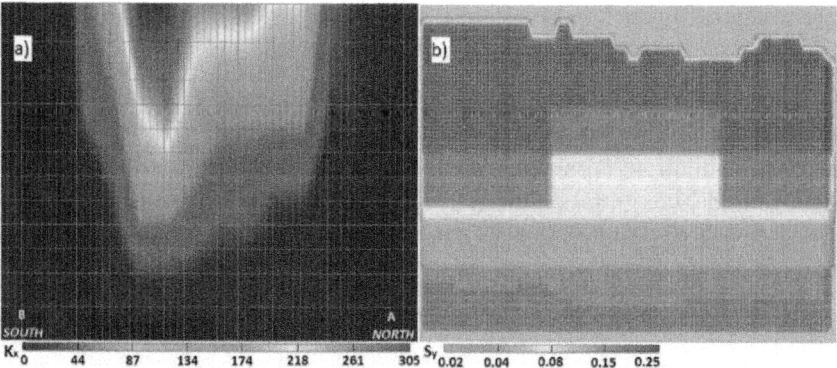

Figure 5: Variability of aquifer hydraulic parameters. (a) Vertical distribution of calibrated hydraulic conductivity (m/day) along cross section A-B (the plot is 50 times ex-

aggerated in the vertical axis); and (b) specific yield of all model layers. doi:10.1371/journal.pone.0138954.g005

The same model parameters (hydraulic conductivity and recharge) used for steady-state calibration were used for the transient calibration to adjust the storage parameters. In the north of the karst upland plateau, with alluvial aquifer and beach deposits, transient calibration resulted in a specific yield (S_y) of 0.25. The limestone aquifers of the karst upland plateau were calibrated with a S_y of 0.08; to the north, west and east of the plateau, where it transitions to the alluvial aquifer, all limestone units were calibrated with S_y of 0.15. The south of the karst upland plateau with clay lenses was calibrated with S_y of 0.02 after a transition area with S_y of 0.04 (c.f., Fig 5B).

A relatively sharp interface between fresh and saline water was estimated for the freshwater lens with a depth to the interface about 40 times of height of freshwater and the saltwater zone can extend up to 7 km inland of the Atlantic Ocean coastline [24]. The thickest part of the freshwater lens (exceeding 100 m) generally lies along a line where the interface with an approximate slope of 2% intersects the base of the aquifer (i.e. southern Cibao Formation).

Commonly, the frequency distribution of the hydraulic conductivity is closer to log-normal than normal [49] in karst aquifers, but could be positively or negatively skewed [29]. The distribution of fracture aperture widths is generally assumed to follow a log-normal [50] or power-law distribution. Conduits spacing should therefore also follow a similar distribution, however, if present, a single conduit can dominate an aquifer [51]. The frequency distribution of the hydraulic conductivity of the present study is presented in Fig 6. Hydrogeological data sets often display a positive skewed frequency distribution, as hydraulic conductivity did in the present study (Fig 6A). The EPM favors a right-skewed distribution toward higher equivalent conductivity values, trying to approximate the individual highly-permeable pathways with an equivalent higher conductivity value in the model cells, which can be better seen in the logarithmic frequency distribution log(K) (Fig 6B). However, one should be aware that these plots do not represent actual field conditions with a possible conduit network, which can impose higher log(K) and potentially result in a log-normal distribution.

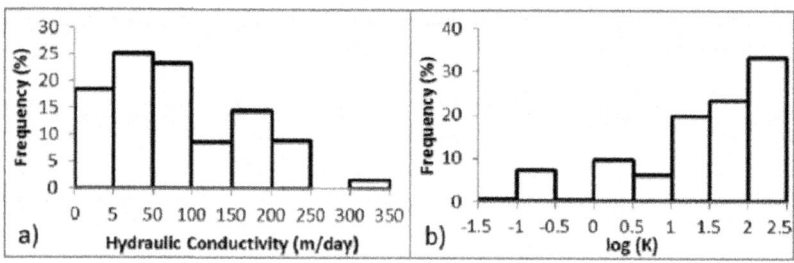

Figure 6: Frequency distribution of calibrated hydraulic conductivity. (a) Calibrated hydraulic conductivity is positively skewed; and (b) Logarithmic hydraulic conductivity log(K) is negatively skewed. doi:10.1371/journal.pone.0138954.g006

The transient transport model was run using monthly rainfall between January 1990 and January 2011. For the initial conditions, the transport model uses the head distribution in January 1990 computed by the flow model, and the TCE distribution in January 1990, the latter generated by linear interpolation of TCE concentration data measured in wells of the contaminated area [46]. Except for the TCE source cell near the eastern part of the Industrial Park (c.f., Fig 3) in the uppermost model layer, the TCE recharge concentration (C_{RCH}) in the study area was set to zero. The TCE mass present as free phase in the source zone or adsorbed onto the media provides ongoing sources of dissolved contamination in the aqueous phase. The mixing of the dissolved TCE in the source zone is a function of the local groundwater flow field within the TCE entrapment zone. Although the influx concentration may vary depending on hydrological conditions, it is considered constant in the transport simulations. The subsequent development of the dissolved TCE plume is controlled by the regional groundwater flow field and various processes such as advection, dispersion, diffusion, adsorption, and a minimal decay.

Since chlorinated organic compounds tend to decay very slowly in groundwater, TCE was assumed conservative with a minimal first-order decay constant of 1.95×10^{-6} d^{-1} (a half-life of 975 years). A linear distribution coefficient (k_d) of 0.2 l/kg (corresponding to a retardation factor of 3.95) was used in this study based on theoretical methods and field measurements on distribution coefficient for sorption of TCE in soil water systems with different organic carbon content [52].

Manual calibration of the transport model was performed based on comparing simulated and measured TCE concentrations between January 1990 and January 2008. Groundwater samples were collected quarterly from monitoring wells in the focus area by USGS and EPA. The environmental and hydrogeologic data were collected from recent studies and US Geological

Survey [34, 46]. All data were compiled into a sustainable database in the Puerto Rico Testsite for Exploring Contamination Threats (PROTECT) Center in Northeastern University for current and future study.

Model Evaluation

To compare the closeness of modeled values to measured values, the model's estimation errors were calculated using two measures of root mean square error (RMSE) and mean absolute error (MAE). Also linear regression analysis was used to evaluate the dependency of water level fluctuations to precipitation events.

RESULTS AND DISCUSSION

One of the main objectives of the study is to evaluate the performance and applicability of the EPM approach to simulate flow and transport for a highly permeable karst aquifer. For this reason, efforts have been undertaken to challenge model results and thoroughly evaluate them against given data from the study site.

Flow Model

The steady-state calibration of the groundwater flow model was accomplished based on minimizing the MAE and RMSE of the simulated potentiometric surface map (Fig 4). Using average conditions in February 1983, RMSE of the model is 1.0889 m and MAE is 0.5230 m for 70 steady-state calibration wells (Fig 7A), representing about 1.7% and 3.6% of the measured water level variations across this unconfined aquifer. The number of available measured heads for transient calibration were much fewer than steady-state data, and the only three water level stations used for transient model calibration (operated by USGS Caribbean Water Science Center) are Dorado Beach #7, and Sabana Hoyos #2 and Bechtel #6 multiport wells (see Fig 3). The long term transient calibration results are presented in Fig 7B as a scatter plot diagram. Based on a statistical analysis of transient model results, the calculated low errors of 0.2049 m (MAE) and 0.2673 m (RMSE) indicate satisfactory simulation of water table variations in those stations. The linear trend line of data ($R^2 =$ 0.6867) is slightly misaligned from the expected 45° line. A Pearson correlation coefficient (R) of 0.8186 between simulated and observed hydraulic heads indicates generally good temporal correlation between observed and simulated water tables. Taking into consideration the used equivalent porous media approach, these values suggest satisfying results for the model calibration. Higher values of hydraulic head are slightly underestimated and lower heads are slightly overestimated by the model. This suggests that the recharge in dry seasons is overestimated, while it is underestimated in wet seasons.

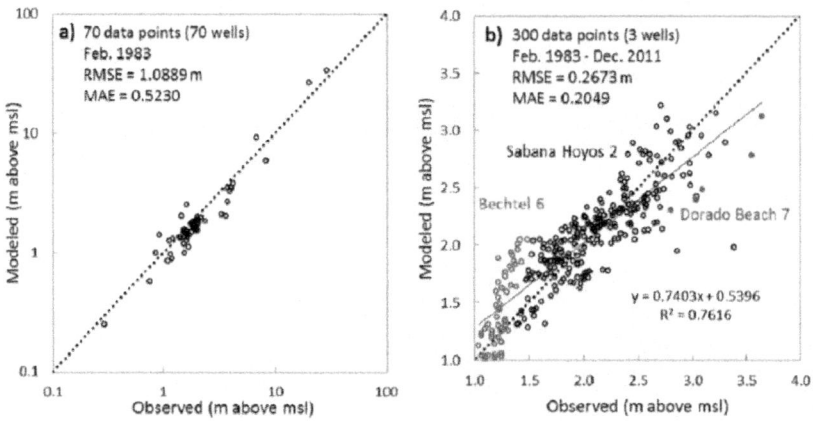

Figure 7: Scatter calibration diagrams of simulated versus observed hydraulic heads. (a) Steady-state calibration of the flow model (70 monitoring wells); and (b) Transient calibration of the flow model over three decades (3 monitoring wells). doi:10.1371/journal.pone.0138954.g007

The simulated steady-state water balance in the aquifer demonstrates that the total aquifer recharge mostly originates from precipitation (98.69%) and river infiltration (1.31%). Groundwater flow in the study area discharges mostly into the Atlantic Ocean (19.62%) in the uppermost two layers, and extraction wells (74.41%). Except the two fourth order springs (together discharging 1.05%), no major springs have been identified in the Vega Alta area (Fig 3), and surface rivers drain only 4.92% of the total groundwater discharge. As net recharge rates increased (decreased) or pumping rate decreased (increased) during the transient simulations, discharge rates to the rivers and to the Atlantic Ocean increased (decreased). Due to absence of natural drainage features (such as wetlands or lagoons); in Vega Alta area more groundwater is available for withdrawals than in adjacent aquifers [53]. As a result, the impact of excessive pumping from this productive unconfined aquifer on water table elevations and flow patterns are significant. The head contours are locally deviated around highly productive wells and the steady-state flow is diverged eastward close to Rio de la Plata River (see Fig 4). During extreme rainfall events, the surface rivers discharge to aquifer because the hydraulic gradient towards the aquifer increases.

As the distance from the coast increases (>6700 m), the Aguada formation at the bottom of the aquifer becomes shallower and reduced in thickness.

As a result, the depth-averaged storativity and conductivity values decrease and the average groundwater level rises faster relative to sea level (Fig 8). The magnitude of seasonal fluctuations depends upon the distance from the constant head boundary condition in the north. The low hydraulic gradient in the central part of the model reflects high hydraulic conductivities (c.f., Fig 5B), while slightly steeper hydraulic gradients reflect the relatively lower hydraulic conductivity in the northern model part. The steepest hydraulic gradient in south reflects the lowest values of hydraulic conductivity in the Vega Alta aquifer, which is a typical relationship between different zones in karst aquifers [5]. The water-table gradient is almost flat (0.57 to 0.76 meters per kilometer) in the Aymamon limestone. It increases to the south in the Aguada limestone (2.85 to 3.79 meters per kilometer) toward the outcrop areas of the Cibao formation (model boundary).

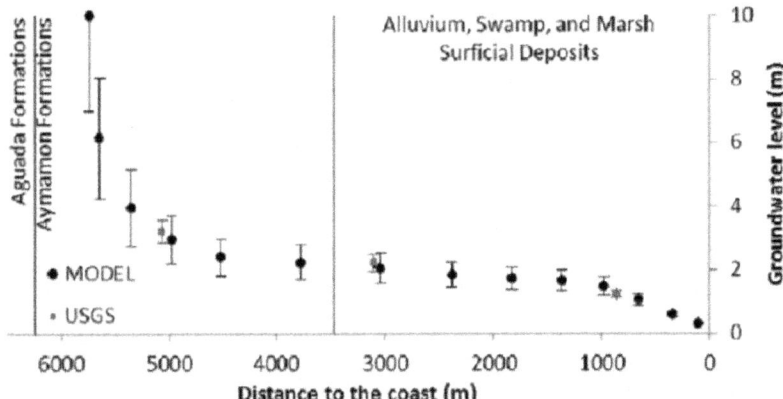

Figure 8: Simulated and measured groundwater levels vs. variable distance to coast. The mean and standard deviation for water level fluctuations in 3 USGS monitoring wells (red) and simulated points along the cross section A-B (see Fig 1). Distance dividers indicate the limit of surficial deposits and the Aguada-Aymamon intersection. doi:10.1371/journal.pone.0138954.g008

Fig 9 shows the long term variation of the water table at three USGS wells and the water table rise of about 0.76 m at Sabana Hoyos #2 well over the last three decades. The rise is due to greater mean annual rainfall and the shutdown of some of the extraction. The simulated transient heads agree well with observed values, validating the assumption of constant withdrawal rates. This also suggests the reliability of EPM approach for replicating annual groundwater level fluctuations in northern Puerto Rico karst system, which agrees well with similar studies [5, 7, 12, 13] but provides validation in higher temporal resolutions (i.e. monthly and daily). Intermediate-term (several years) variation in water table corresponds to wet years, dry years and groundwater withdrawal activities. After 1967, the next severe dry period of the twentieth century in Puerto Rico was from 1993 to 1995 [54], when the annual rainfall accumulation was below normal with the greatest rainfall deficit.

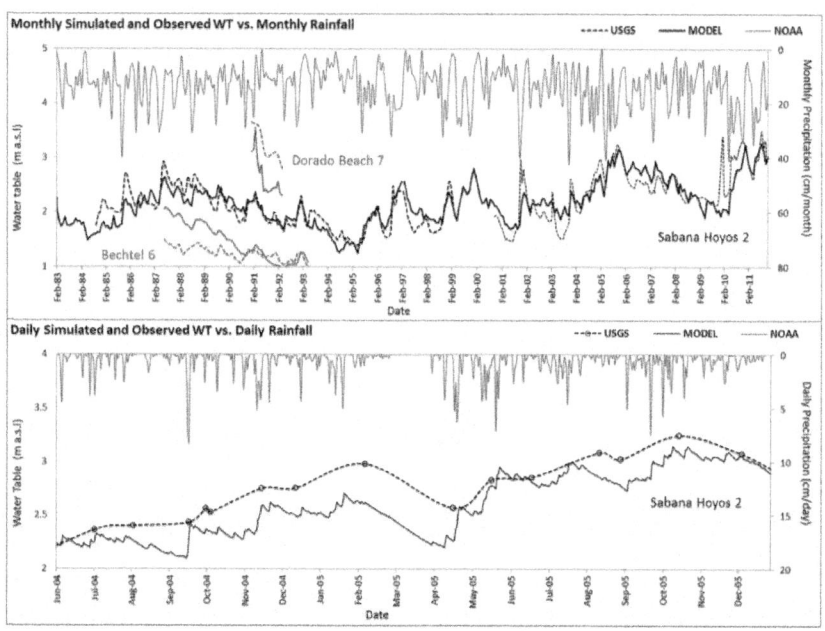

Figure 9: Monthly and daily averaged hydrogeologic conditions in Vega Ata aquifer. Simulated (solid line) and observed (dotted line) monthly water level hydrographs of three USGS wells vs. precipitation between 1983–2011 (upper); simulated and observed daily water levels in one well between June 2004 –December 2005 (lower). doi:10.1371/journal.pone.0138954.g009

Precipitation is often higher in winter than in summer. Therefore, the karst storage may not be fully recharged in summer and can demonstrate seasonal water table variations. The study results reveal a strong relationship between rainfall and water table fluctuations, demonstrating that the groundwater table was mostly recharged by the rainfall. In addition to climatic conditions, water table fluctuation also responds to the closing of contaminated public water supply wells and the operation of new ones between February 1983 and April 1992, after which there have not been more wells reported as closed. The results illustrate that the water table level achieved its overall minimum in 1995 and maximum in 2011.

Monthly rainfall intensities varied between 0 and 42.67 cm/month, with an average intensity of about 14.73 cm/month. The water table responds soon after the occurrence of intense rainfalls which suggests rapid infiltration and limited delayed recharge through vadose zone. The match between the observed and simulated response of the water table to rainfall events can be well reproduced by the model (Fig 9). For example, short-term seasonal groundwater levels in 2005 varied up to a maximum of 0.6 m according to both measured and simulated data (Fig 9). Analyses of different recharge events suggest that aquifer replenishment in the model responds greatly to rainfalls of any magnitude. It can be observed in the model results that the increasing response to short-term precipitation events, such as storms, is faster than the groundwater decline following the storm.

Water table recovery, which is important in aquifer protection and sustainable water resource management, is affected by factors including precipitation rate, recharge and discharge rates, porosity, topography, and geologic structure. The water table recovery after each precipitation event was faster in the wet season and is observed several hours or several days after rainfall events in the study area. The linear correlation of monthly rainfall intensities with the water table recovery observed by Chuanmao [55] is acknowledged by the model results of the present study in the Vega Alta aquifer. It is thus possible to estimate the water table raise in each month based on monthly rainfall intensity. For example, the linear regression equation for water table recovery at Sabana Hoyos #2 well ($R^2 = 0.9132$) is shown on Fig 10, where ΔWT is monthly recovery of water table (m), and P is the monthly precipitation (cm/month).

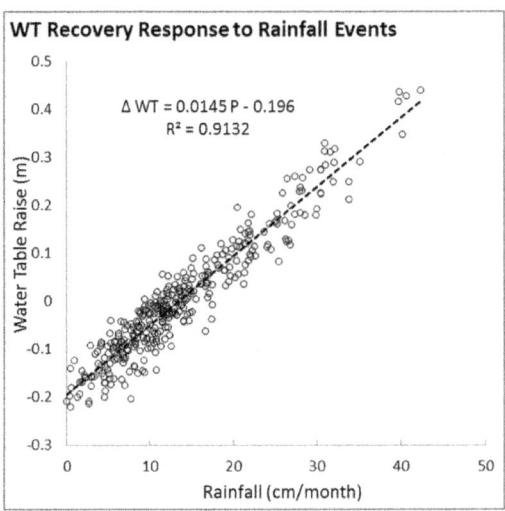

Figure 10: Scatter diagram of simulated water level recovery as a linear function of rainfall. On average, a monthly rainfall less (more) than 13.5 cm would

lower (increase) the water table in USGS Sabana Hoyos #2 well. doi:10.1371/journal.pone.0138954.g010

The hydrogeologic effects of subsurface conduit networks cannot be reproduced using an EPM modeling approach, nor can be well detected using limited number of head monitoring wells in this study. The developed flow model for Vega Alta karst aquifer adequately simulates the spatial distribution and high-resolution temporal variation of groundwater levels over 21 years. It also uses monthly-refined rainfall, and captures the water table fluctuations in a larger temporal resolution for a shorter time period using daily-refined rainfall. Other flow modeling studies in northern Puerto Rico (e.g. Sepúlveda, 1996; Cherry, 2001) only demonstrated EPM's ability to regenerate steady-state regional hydrogeology, or transient simulations using very coarse rainfall resolution (e.g. annual).

Contaminant Transport Model

The calibration was accomplished by adjusting C_{RCH} value to 1,700 µg/l in the TCE source cell. Based on this value, the average TCE influx from the source area can be calculated by multiplying C_{RCH} with the recharge rate and the TCE source cell size, resulting in 7.035 kg/yr TCE influx into the unconfined aquifer. To control the mixing of the TCE in the water, uniform scale-dependent longitudinal, transverse, and vertical dispersivity values of 3, 0.3, and 0.03 m, respectively, were resulted during the calibration for all model layers, with assuming a typical molecular diffusion coefficient k_m of 10^{-5} cm^2/s. The calibrated model agrees well with observations in 10 monitoring wells in the area, and the correlation coefficient between simulated and measured TCE concentrations is 0.857 (Fig 11). Besides the limited field data used, the model simulation of concentrations produced an overall RMSE of 59.10 µg/l (normalized RMSE = 8.45%). This correlation suggests that the use of an EPM approach to simulate solute transport in an intermediate scale for this eogenetic karst limestone aquifer is justified. As noted by other studies [5], EPMs are not suitable for transport modeling in a local scale, where in-depth knowledge of distribution patterns of fracture and conduit networks is required to describe the complex transport processes.

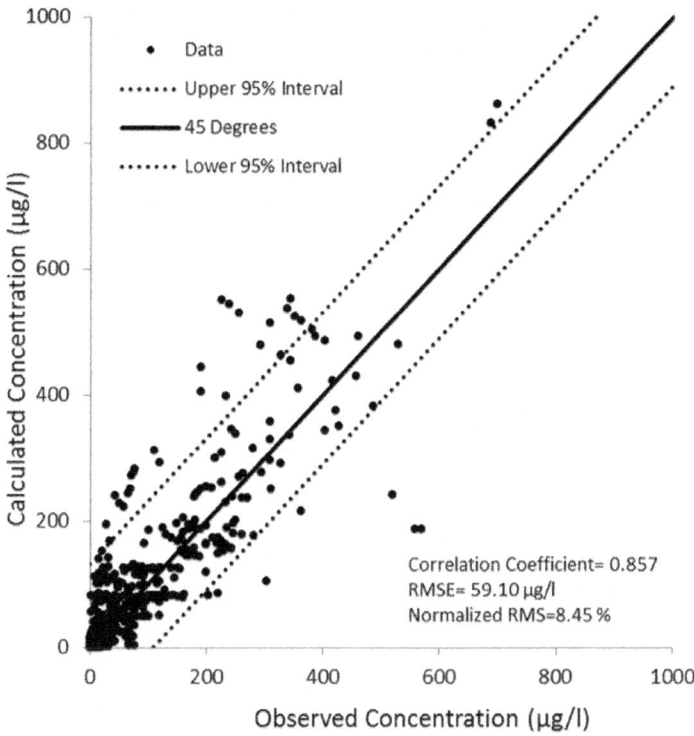

Figure 11: Transient calibration diagram for the transport model. Scatter diagram of simulated vs. observed TCE concentrations at various depths in 10 monitoring wells between January 1990 and January 2008. doi:10.1371/journal.pone.0138954.g011

TCE concentrations were measured at several observation wells in discrete sampling zones at different depths with the deepest at -82.6 m a.s.l. Percentile curve for the simulated TCE concentrations show good agreement with that for the TCE measurements distributed over the focus area (Fig 12). High degrees of spatial and vertical variation in TCE concentrations indicate the necessity in three dimensional simulating of the complex karst environment. The zone of high TCE concentrations extends to the depth of the deepest wells in the area (-61 m a.s.l).

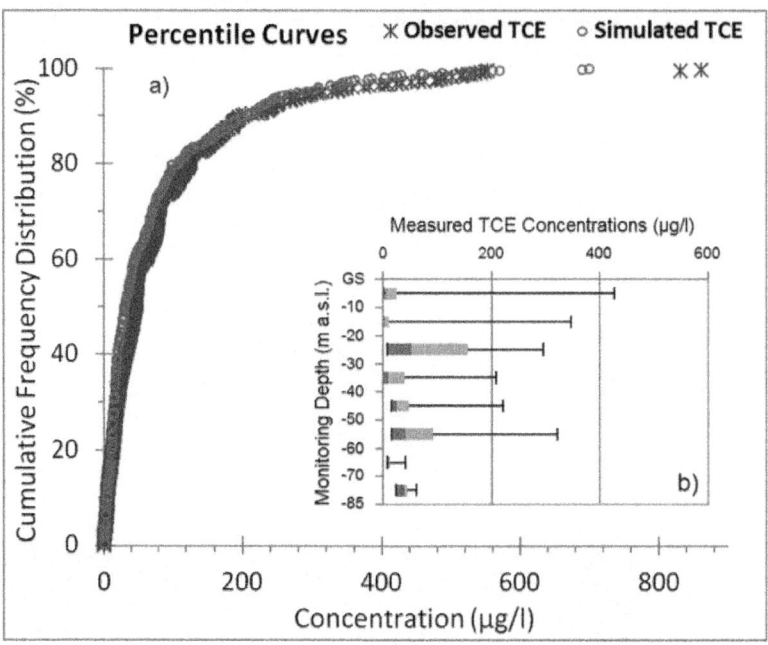

Figure 12: Cumulative frequency and vertical distribution of detected TCEs in ground-water. (a) Cumulative quantile plots of observed and simulated TCE concentrations were obtained from historical data and model simulations, respectively. (b) Box plots for the depth distribution of observed TCE concentrations indicate 5%, 25%, 50%, 75%, and 95% quantiles (GS: ground surface). doi:10.1371/journal.pone.0138954. g012

In northern Puerto Rico, 56.6% of samples had TCE concentrations above detection limit (DL) and 11.5% of those samples had TCE concentrations above maximum contamination level (MCL) of 5 µg/l established by EPA. The mean, median, standard deviation, and coefficients of variation (CV %) for all measured TCE concentrations were 83.7 µg/l, 46.3 µg/l, 112.6 µg/l, and 41.1%, respectively, indicating that concentrations were substantially skewed (γ: 2.78) toward lower values. The sampling frequency and distribution have decreased significantly over recent years causing spatial uncertainties [56, 57]. Historically no significant seasonal fluctuation of TCE concentrations was observed in the taken samples (Yu et al., 2015); therefore in 2007, EPA proposed to reduce the sampling frequency from quarterly to semi-annually or annually. No TCE concentration was reported for Mackovic Spring in Vega Alta or Maguayo Spring in Dorado, both located outside of the focus area. Variability of spatially detected concentrations is commonly associated with complex subsurface hydrodynamics that affect storage and mobility of contaminants. The initial distribution of TCE concentration and the simulated spreading of TCE plume in the focus area at a depth of -30.48 m a.s.l between 1993 and 2008 are presented in Fig 13. The presented observation data are interpolated values collected from monitoring zones between depths of -25.91 to -36.58 m a.s.l. The spreading of the plume is simply observed by the outermost concentration contour which corresponds to the MCL (5 μg/l). The dominant flow path in the focus area is slightly northeast and downward towards the deeper aquifer, and concentrations directly south of the Industrial Park are negligible.

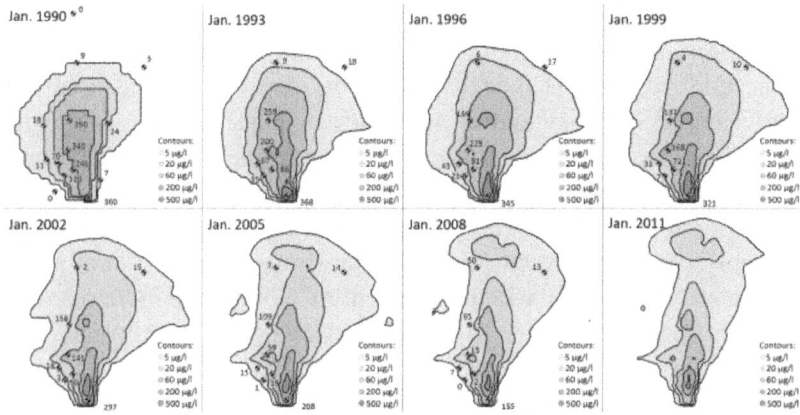

Figure 13: The migration and extent of TCE plume over two decades. Initial TCE contours were based on available data at January 1990, and simulated TCE contours and observed TCE data (unit: μg/l) were at eight monitoring wells at a depth of -30.48 m a.s.l between January 1993 and January 2011 in the focus area (4.6 × 4.8 km, see Fig 1). No observation data was available for 2011. doi:10.1371/journal.pone.0138954. g013

Due to the very low transport velocities, the TCE plume is very unlikely to reach the Atlantic Ocean. Sensitivity analysis suggests that the plume would spread further downward and to the north if no sorption of TCE was accounted for. The vertical hydraulic gradient becomes smaller toward deeper parts of the aquifer and toward the Atlantic Ocean. Accordingly, the simulation indicates a slightly downward and eastward transport in addition to the general northward movement of the TCE plume especially in the shallower parts of the aquifer.

The closure of some public water supply wells on the right side of focus area until 1992 has pushed the slight northeast groundwater direction further to the east. The unknown preferential flow paths and complex network of fractures, conduits and channels in this karst aquifer causes irregular spreading patterns and affects the contaminant dispersal and storage mechanisms. Measured TCE concentrations in different wells within the focus area had different temporal patterns reflecting high subsurface anisotropy and heterogeneity and different attenuation rates (also see Yu et al., 2015). In some wells, high variations of TCE concentrations were observed between consecutive monitoring events that are not captured by the model, e.g. after the plume had progressed enough (2005), the TCE concentrations in one site north of the plume started to increase gradually over time which was not seen by the model. The highest TCE concentration occurs directly downstream of the source cell and as the distance to the source increases, generally the TCE concentration decreases.

There are notable decreasing trends in the values and areas of TCE contamination on the temporal scale, as supported by the recent measured data. The total mass of TCE in Vega Alta aquifer was estimated as 5889.4 kg (equivalent to TCE volume of 4.033 m^3) in January 1990. Although the annual TCE influx (7.035 kg) was contributing to the subsurface plume, the total contamination reduced to estimated total mass of 2501.4 kg in January 2011. Besides sorption and natural attenuation, this is partially because public water supply wells (before closure) and industrial and agricultural pumping wells withdrew the aquifer within the extent of TCE plume in the focus area. The TCE removal rate is found to be higher at shallower layers where vertical hydraulic gradient is larger.

Freshwater resources within the northern karst systems of Puerto Rico are highly vulnerable to contamination. Especially the recent industrial and urban development and increased number of contaminated sites within the island lead to unintended contamination of the groundwater resources. The subsequent long-term contamination indicates a large capacity of the karst aquifers for storing and releasing contaminants which reflects potential lasting exposures.

CONCLUSIONS

This paper evaluates the application and adequacy of EPM approach to simulate groundwater flow and contaminant fate and transport processes within highly permeable karst environments, which is crucial for managing water resources management purposes. Where little to nothing is known about the occurrence and properties of conduit and fracture networks, approximating the complex nature of karst systems with equivalent hydraulic properties may sufficiently reproduce the aquifer behavior at regional and intermediate scales. Locating those networks and identifying the associated hydrogeological parameters are impossible without large scale exploration efforts. Using limited field data, the developed MODFLOW and MT3D models in this study adequately simulated the spatial distribution and high-resolution temporal variation of groundwater levels and the decreasing trend of TCE plume over 21 years, respectively, for Vega Alta karst aquifer. Withdrawals from a large well field, as the main regional discharge feature, are evaluated to significantly impact the water table dynamics and the dominant direction of plume migration. The occurrence of preferential flow influences the flow conditions and leads to irregular spreading of TCE plume in the aquifer. The main limitations of equivalent porous media approach in karst aquifers are failing to capture groundwater hydrodynamics on a local scale, and to simulate transient turbulent flow through conduit network and its interaction with the matrix.

ACKNOWLEDGMENTS

We thank Christine Gordon for her editorial help on the paper. We also thank the editor and the anonymous reviewers in advance for their valuable comments.

AUTHOR CONTRIBUTIONS

Conceived and designed the experiments: RG XY CB FLH IYP ANA. Performed the experiments: RG. Analyzed the data: RG XY CB FLH. Contributed reagents/materials/analysis tools: RG XY CB FLH IYP ANA. Wrote the paper: RG XY CB FLH. Initiated this work: IYP ANA.

REFERENCES

1.   Ford DC, Williams PW (2007) Karst hydrogeology and geomorphology. John Wiley & Sons 576.

2.   Butscher C, Auckenthaler A, Scheidler S, Huggenberger P (2011) Validation of a numerical indicator of microbial contamination for karst springs. Ground Water 49(1):66–76. doi: 10.1111/j.1745-6584.2010.00687.x. pmid:20180864

3.   Chen Z, Goldscheider N (2014) Modeling spatially and temporally varied hydraulic behavior of a folded karst system with dominant conduit drainage at catchment scale, Hochifen-Gottesacker, Alps. Journal of Hydrology 514:41–52.

4.   Liedl R, Sauter M, Huckinghaus D, Clemens T, Teutsch G (2003) Simulation of the development of karst aquifers using a coupled continuum pipe flow model. Water Resources Research 39(3):1057. doi: 10.1029/2001wr001206

5.   Scanlon BR, Mace RE, Barrett ME, Smith B (2003) Can we simulate regional groundwater flow in a karst system using equivalent porous media models? Case study, Barton Springs Edwards aquifer, USA.

Journal of Hydrology 276:137–158. doi: 10.1016/s0022-1694(03)00064-7

6.   Worthington SRH (2009) Diagnostic hydrogeologic characteristics of a karst aquifer (Kentucky, USA). Hydrogeology Journal 17:1665–1678. doi: 10.1007/s10040-009-0489-0

7.   Abusaada M, Sauter M (2013) Studying the Flow Dynamics of a Karst Aquifer System with an Equivalent Porous Medium Model. Groundwater 51:641–650. doi: 10.1111/j.1745-6584.2012.01003.x

8.   Bodin J, Ackerer P, Boisson A, Bourbiaux B, Bruel D, Dreuzy JR de, et al. (2012) Predictive modelling of hydraulic head responses to dipole flow experiments in a fractured/karstified limestone aquifer: Insights from a comparison of five modelling approaches to real-field experiments. Journal of Hydrology 454–455. doi: 10.1016/j.jhydrol.2012.05.069

9.   Ghasemizadeh R, Hellweger F, Butscher C, Padilla I, Vesper D, Field M, et al.(2012) Review: Groundwater flow and transport modeling of karst aquifers, with particular reference to the north coast limestone aquifer system of Puerto Rico. Hydrogeology Journal 20:1441–1461. pmid:23645996 doi: 10.1007/s10040-012-0897-4

10.   Panagopoulos G (2012) Application of MODFLOW for simulating groundwater flow in the Trifilia karst aquifer, Greece. Environmental Earth Sciences 67:1877–1889. doi: 10.1007/s12665-012-1630-2

11.   Anderson MP, Woessner WW (1992) Applied groundwater modeling: simulation of flow and advective transport. Academic Press, New York, 381p.

12.   Long JCS, Remer JS, Wilson CR, Witherspoon PA (1982) Porous Media Equivalents for Networks of Discontinuous Fractures. Water Resources Research 18(3):645–658. doi: 10.1029/wr018i003p00645

13.   Pankow JF, Johnson RL, Hewetson JP, Cherry JA (1986) An evaluation of contaminant migration patterns at two waste disposal sites on fractured porous media in terms of the equivalent porous medium (EPM) model. J. Contaminant Hydrology 1(1):65–76. doi: 10.1016/0169-7722(86)90007-0

14.   Hartmann A, Goldscheider N, Wagener T, Lange J, Weiler M (2014) Karst water resources in a changing world: Review of hydrological modeling approaches. Reviews of Geophysics, 52(3):218–242. doi: 10.1002/2013rg000443

15.   Putnam LD, Long AJ (2009) Numerical groundwater-flow model of the Minnelusa and Madison hydrogeologic units in the Rapid City area, South Dakota. U.S. Geological Survey Scientific Investigations Report

2009–5205.

16. Saller SP, Ronayne MJ, Long AJ (2013) Comparison of a karst groundwater model with and without discrete conduit flow. Hydrogeology Journal 21(7):1555–1566. doi: 10.1007/s10040-013-1036-6

17. Quinn J, Tomasko DA (2000) Numerical Approach to Simulating Mixed Flow in Karst Aquifers. Groundwater Flow and Contaminant Transport in Carbonate Aquifers 147–156.

18. Quinn J, Tomasko D, Kuiper JA (2006) Modeling complex flow in a karst aquifer. Sedimentary Geology 184(3):343–351. doi: 10.1016/j.sedgeo.2005.11.009

19. Lindgren RJ, Dutton AR, Hovorka SD, Worthington SRH, Painter S (2004) Conceptualization and simulation of the Edwards Aquifer, San Antonio Region, Texas. U.S. Geological Survey Scientific Investigations Report 2004–5277.

20. McDonald MG, Harbaugh AW (1988) A modular three-dimensional finite-difference ground-water flow model. U.S. Geological Survey Report 83–875.

21. Zheng C, Wang PP (1999) MT3DMS: A Modular Three-Dimensional Multispecies Transport Model for Simulation of Advection, Dispersion, and Chemical Reactions of Contaminants in Groundwater Systems; Documentation and User's Guide. U.S. Army Engineer Research and Development Center Contract Report SERDP-99-1; Vicksburg, Mississippi.

22. Lugo AE, Castro LM, Vale A, del Mar Lopez T, Prieto EH, Martino AG (2001) Puerto Rican karst—A vital resource. U.S. Department of Agriculture and Forest Service General Technical Report WO-65.

23. Molina-Rivera WL, Gómez-Gómez F (2008) Estimated water use in Puerto Rico, 2005: 454 U.S. Geological Survey Open-File Report 2008–1286.

24. Renken RA, Ward WC, Gill IP, Gómez GF, Rodriguez JM (2002) Geology and Hydrogeology of the Caribbean Islands Aquifer System of the Commonwealth of Puerto Rico and the US Virgin Islands. U.S. Geological Survey Professional Paper 1419:1–139.

25. Ginés A, Ginés J (2007) Eogenetic karst, glacioeustatic cave pools and anchialine environments on Mallorca Island: a discussion of coastal speleogenesis. International Journal of Speleology 36(2):57–67. doi: 10.5038/1827-806x.36.2.1

26. Padilla I (2013) Fate of DNAPLs in eogenetic karst aquifers: the case of

northern Puerto Rico. 2013 GSA Annual Meeting, Denver CO.

27. Anaya A, Padilla I, Macchiavelli R, Vesper D, Meeker J, Alshawabkeh A (2014) Estimating preferential flow in karstic aquifers using statistical mixed models. Groundwater 52(4):584–596. doi: 10.1111/gwat.12084

28. Gulley JD, Martin JB, Moore PJ, Brown A, Spellman PD, Ezell J (2015) Heterogeneous distributions of CO_2 may be more important for dissolution and karstification in coastal eogenetic limestone than mixing dissolution. Earth Surface Processes and Landforms. doi: 10.1002/esp.3705

29. Budd DA, Vacher HL (2004) Matrix permeability of the confined Floridan Aquifer, Florida, USA. Hydrogeology Journal 12:531–549. doi: 10.1007/s10040-004-0341-5

30. Monroe WH (1963) Geology of the Vega Alta Quadrangle, Puerto Rico, U.S. Geological Survey Geologic Quadrangle Map GQ-191.

31. Monroe WH (1980) Geology of the middle Tertiary formations of Puerto Rico. U.S. Geological Survey Professional Paper 953.

32. Giusti EV, Bennett GD (1976) Water resources of the North Coast limestone area. U.S. Geological Survey Water-Resources Investigations Report 42–75.

33. Giusti EV (1978) Hydrogeology of the karst of Puerto Rico. U.S. Geological Survey professional paper 1012. 68 p.

34. Sepúlveda N (1996) Three-dimensional ground-water-flow model of the water-table aquifer in Vega Alta, Puerto Rico. U.S. Geological Survey. Water-Resources Investigations Report 95–4184.

35. Rodriguez JM (1995) Hydrology of the North Coast limestone aquifer system in Puerto Rico. U.S. Geological Survey Water Resources Report 94–4249.

36. Rodriguez JM, Hartley JL (1994) Geologic and Hydrologic Data Collected at Test Holes NC-6 and NC-11, Hatillo and Isabela, Northwestern Puerto Rico. U.S. Department of the Interior, U.S. Geological Survey 93–465.

37. Rodriguez JM, Richards RT (2000) Detection of conduit-controlled ground-water flow in northwestern Puerto Rico using aerial photograph interpretation and geophysical methods. U.S. Geological Survey Water Resource Investigation Report 00–4147.

38. US Environmental Protection Agency (1997) Superfund Record of Decision (EPA Region 2): Vega Alta Public Supply Wells, Operable Unit 2, Vega Alta PR. OCLC Number: 48664598, Stock Number: PB97-963821; EPA ID: PRD980763775.

39. Rodriguez MJ (1997) Characterization of springflow in the North Coast limestone of Puerto Rico using physical, chemical, and stable isotopic methods. U.S. Geological Survey Water Resources Investigations Report 97–4122.

40. Veve TD, Taggart BE (1996) Atlas of groundwater resources in Puerto Rico and the US Virgin Islands. U.S. Geological Survey Water Resource Investigation Report 94–4198.

41. Gómez GF, Torres-Sierra H (1988) Hydrology and effects of development on the water-table aquifer in the Vega Alta quadrangle. U.S. Geological Survey Water-Resources Investigations Report 87–4105.

42. Guzmán RS, García R, Avilés A (1986) Reconnaissance of Volatile Synthetic Organic Chemicals at Public Water Supply Wells throughout Puerto Rico, November 1984–May, 1985. U.S. Geological Survey. Open-File Report 86–63.

43. .Zack A, Rodriguez-Alonso T, Romas-Mas A (1987) Puerto Rico Groundwater Quality. U.S. Geological Survey. Open File Report 87–0749.

44. Guzmán RS, Quiñones MF (1984) Ground-water quality at selected sites throughout Puerto Rico, September 1982-July 1983. U.S. Geological Survey. Open-File Rep 84–05.

45. Schlumberger (2011) Visual MODFLOW 2011.1 User's Manual. Schlumberger Water Services (SWS). Kitchener, ON, Canada.

46. Sepúlveda N (1999) Ground-water flow, solute transport, and simulation of remedial alternatives for the water-table aquifer in Vega Alta, Puerto Rico. U.S. Geological Survey. Water-Resources Investigations Report 97–4170.

47. Gómez-Gómez F (2008) Estimation of the Change in Freshwater Volume in the North Coast Limestone Upper Aquifer of Puerto Rico in the Río Grande de Manatí-Río de la Plata Area between 1960 and 1990 and Implications on Public-Supply Water Availability. U.S. Geological Survey. Scientific Investigations Report 2007–5194.

48. Torres-Sierra H (1985) Potentiometric surface of the upper limestone aquifer in the Dorado-Vega Alta area, north-central Puerto Rico, February 1983. U.S. Geological Survey Water-Resources Investigations Report 85–4268.

49. Halihan T, Mace RE, Sharp JM (2000) Flow in the San Antonio segment of the Edwards aquifer: matrix, fractures, or conduits? Groundwater flow and contaminant transport in carbonate aquifers Wicks and Sasowsky (eds), AA Balkema, Rotterdam, Netherlands 129–146.

50. Romanov D, Gabrovšek F, Dreybrodt W (2003) Dam sites in soluble rocks: a model of increasing leakage by dissolutional widening of fractures beneath a dam. Engineering Geology 70(1):17–35. doi: 10.1016/s0013-7952(03)00073-5

51. Halihan T, Sharp JM, Mace RE (1999) Interpreting flow using permeability at multiple scales. Karst Modeling: Proceedings of the symposium Charlottesville, VA. Palmer, Palmer, and Sasowsky (eds). Special Publication 5, Karst Waters Institute 82–96.

52. Mehran M, Olsen RL, Rector BM (1987) Distribution Coefficient of Trichloroethylene in Soil Water Systems. Ground Water 25:275–282. doi: 10.1111/j.1745-6584.1987.tb02131.x

53. Cherry GS (2001) Simulation of flow in the upper North Coast limestone aquifer, Manatí-Vega Baja area, Puerto Rico. U.S. Geological Survey Water Resources Investigation Report 00–4266.

54. Larsen MC (2000) Analysis of 20th century rainfall and streamflow to characterize drought and water resources in Puerto Rico. Physical Geography 21(6):494–521.

55. Chuanmao J (1982) Variation of the groundwater regime under the effects of human activities and its artificial control Improvements of Methods of Long Term Prediction of Variations in Groundwater Resources and Regimes Due to Human Activity. IAHS Publ No. 136.

56. Yu X, Ghasemizadeh R, Padilla I, Irizarry C, Kaeli D, Alshawabkeh A (2015) Spatiotemporal changes of CVOC concentrations in karst aquifers: Analysis of three decades of data from Puerto Rico. Science of the Total Environment 511:1–10. doi: 10.1016/j.scitotenv.2014.12.031. pmid:25522355

57. Yu X., Ghasemizadeh R., Padilla I., Meeker J. D., Cordero J. F., & Alshawabkeh A (2015) Sociodemographic patterns of household water-use costs in Puerto Rico. Science of the Total Environment 524:300–309. doi: 10.1016/j.scitotenv.2015.04.043. pmid:25897735

CITATION

CHAPTER 1

Xi Zhang and Rob Jeffrey (2013). Development of Fracture Networks Through Hydraulic Fracture Growth in Naturally Fractured Reservoirs, Effective and Sustainable Hydraulic Fracturing, Dr. Rob Jeffrey (Ed.), ISBN: 978-953-51-1137-5, InTech, DOI: 10.5772/56405.

CHAPTER 2

Marcelo Gomes Miguez, Aline Pires Veról and Paulo Roberto Ferreira Carneiro (2012). Sustainable Drainage Systems: An Integrated Approach, Combining Hydraulic Engineering Design, Urban Land Control and River Revitalisation Aspects, Drainage Systems, Prof. Muhammad Salik Javaid (Ed.), ISBN: 978-953-51-0243-4, InTech, DOI: 10.5772/33896.

CHAPTER 3

S. Jung, J. Kang, I. Hong and H. Yeo, "Case Study: Hydraulic Model Experiment to Analyze the Hydraulic Features for Installing Floating Islands," Engineering, Vol. 4 No. 2, 2012, pp. 90-99. doi: 10.4236/eng.2012.42012.

CHAPTER 4

Yang X-D, Zhang X-N, Lv G-H, Ali A (2014) Linking Populus euphraticaHydraulic Redistribution to Diversity Assembly in the Arid Desert Zone of Xinjiang, China. PLoS ONE 9(10): e109071. doi:10.1371/journal.pone.0109071.

CHAPTER 5

Giuliana Zanchi, Salim Belyazid, Cecilia Akselsson, Lin Yu, Kevin Bishop, Stephan J. Köhler and Harald Grip, A Hydrological Concept including Lateral

Water Flow Compatible with the Biogeochemical Model ForSAFE, doi:10.3390/hydrology3010011.

CHAPTER 6

Ronald J. Adrian & Ivan Marusic (2012) Coherent structures in flow over hydraulic engineering surfaces, Journal of Hydraulic Research, 50:5, 451-464, DOI: 10.1080/00221686.2012.729540.

CHAPTER 7

Willi H. Hager & Robert M. Boes (2014) Hydraulic structures: a positive outlook into the future, Journal of Hydraulic Research, 52:3, 299-310, DOI: 10.1080/00221686.2014.923050.

CHAPTER 8

Yongfu Xu (2013). Unsaturated Hydraulic Conductivity of Fractal-Textured Soils, Hydraulic Conductivity, Dr. Vanderlei Rodrigues Da Silva (Ed.), ISBN: 978-953-51-1208-2, InTech, DOI: 10.5772/56716.

CHAPTER 9

Ghasemizadeh R, Yu X, Butscher C, Hellweger F, Padilla I, Alshawabkeh A (2015) Equivalent Porous Media (EPM) Simulation of Groundwater Hydraulics and Contaminant Transport in Karst Aquifers. PLoS ONE 10(9): e0138954. doi:10.1371/journal.pone.0138954

INDEX